Advanced Processing and Manufacturing Technologies

Advanced Processing and Manufacturing Technologies

Contributors

Sylvestre Uwizeyemungu, Placide Poba-Nzaou et al.

AURIS
Reference

www.aurisreference.com

Advanced Processing and Manufacturing Technologies

Contributors: Sylvestre Uwizeyemungu, Placide Poba-Nzaou et al.

Published by Auris Reference Limited
www.aurisreference.com

United Kingdom

Advanced Processing and Manufacturing Technologies

ISBN: 978-1-78154-930-8

British Library Cataloguing in Publication Data
A CIP record for this book is available from the British Library

Printed in the United Kingdom

Exclusively distributed by CBS Publishers & Distributors Pvt. Ltd.

Sales & Distribution Rights only for India, Pakistan, Bangladesh, Sri Lanka, Nepal and Bhutan.This book is not to be sold outside these territories.

Contents

List of Abbreviations .... *vii*

List of Contributors.. ....*ix*

Preface... ....*xiii*

Chapter 1   **Assimilation Patterns in the Use of Advanced Manufacturing Technologies in Smes: Exploring their Effects on Product Innovation Performance**..1

Chapter 2   **Effective Factors on Advanced Manufacturing Technology Implementation Performance: A Review**...23

Chapter 3   **Effects of Computer-Aided Manufacturing Technology on Precision of Clinical Metal-Free Restorations**......................................51

Chapter 4   **A Review of Additive Manufacturing**...63

Chapter 5   **A Novel Manufacturing Technology for Rf Mems Devices on Ceramic Substrates** 85

Chapter 6   **Complex Metallic Alloys as New Materials for Additive Manufacturing** 99

Chapter 7   **A Bilayer Resource Model for Cloud Manufacturing Services** 121

Chapter 8   **Identifying Enablers of E-Manufacturing** 141

Chapter 9   **Biocompatibility of Advanced Manufactured Titanium Implants— A Review**...153

Chapter 10   **Linking Process Technology and Manufacturing Performance Under the Framework of Manufacturing Strategy** 183

Chapter 11   **Advanced Manufacturing Technology Projects Justification** 211

Chapter 12   **Flexible Manufacturing System Simulation Using Petri Nets**............233

Chapter 13   Enhancing Multistage Deep-Drawing and Ironing Manufacturing
              Processes of Axisymmetric Components: Analysis and
              Experimentation..247

              Citations..271

              Index..275

List of Abbreviations

ABS	Acrylonitrile butadiene styrene
AM	Additive manufacturing
AMT	Advanced manufacturing technologies
AGVS	Automated guided vehicles
CD	Cloud emander
CE	Cloud end
CMSM	Cloud manufacturing service model
CP	Cloud provider
CMAs	Complex metallic alloys
CAD	Computer aided design
CIM	Computer integrated manufacturing
CNC	Computer numerical control
CAD	Computer-aided design
CAM	Computer-aided manufacturing
CAPP	Computer-aided process planning
CRM	Customer relationship management
DRES	Distributed real-time embedded systems
EBS	Electron beam sintering
ERP	Enterprise resource planning
FMS	Flexible manufacturing systems
FDA	Food and drug administration
FDM	Fused deposition modeling
IT	Information technologies
ISO	International standards organisation
LENS	Laminated engineered net shaping
LOM	Laminated Object manufacturing
LE	Large enterprises
OA	Office automation
PC	Polycarbonate
RBV	Resource-based view
SEM	Scanning electron microscopy
SLS	Selective laser sintering
SME	Society of manufacturing engineer
SD	Standard deviation
SLC	Stereolithography contour
SCM	Supply chain management
TQM	Total quality management
WSDL	Web service escription language

List of Contributors

Sylvestre Uwizeyemungu
Université du Québec à Trois-Rivières, Trois-Rivières, Canada

Placide Poba-Nzaou
Université du Québec à Montréal, Montréal, Canada

Josée St-Pierre
Université du Québec à Trois-Rivières, Trois-Rivières, Canada

S. Saberi
Department of Mechanical and Manufacturing Engineering, Faculty of Engineering, University Putra Malaysia, 43400 UPM Serdang, Selangor, Malaysia

R. Mohd. Yusuff
Department of Mechanical and Manufacturing Engineering, Faculty of Engineering, University Putra Malaysia, 43400 UPM Serdang, Selangor, Malaysia

N. Zulkifli
Faculty of Engineering, University Pertahanan Nasional Malaysia, Kern Sungai Besi. 57000 Kuala Lumpur, Malaysia

Ki-Hong Lee
Department of Prosthodontics, School of Dentistry and Dental Research Institute, Seoul National University, Daehak-ro, Jongno- gu, Seoul 110-749, Republic of Korea

In-Sung Yeo
Department of Prosthodontics, School of Dentistry and Dental Research Institute, Seoul National University, Daehak-ro, Jongno- gu, Seoul 110-749, Republic of Korea

Benjamin M. Wu
Division of Advanced Prosthodontics, UCLA School of Dentistry, Los Angeles, CA 90095, USA

Jae-Ho Yang
Department of Prosthodontics, School of Dentistry and Dental Research Institute, Seoul National University, Daehak-ro, Jongno- gu, Seoul 110-749, Republic of Korea

Jung-Suk Han
Department of Prosthodontics, School of Dentistry and Dental Research Institute, Seoul National University, Daehak-ro, Jongno- gu, Seoul 110-749, Republic of Korea

Sung-Hun Kim
Department of Prosthodontics, School of Dentistry and Dental Research Institute, Seoul National University, Daehak-ro, Jongno- gu, Seoul 110-749, Republic of Korea

Yang-Jin Yi
Department of Prosthodontics, School of Dentistry and Dental Research Institute, Seoul National University, Daehak-ro, Jongno- gu, Seoul 110-749, Republic of Korea

Taek-Ka Kwon
Department of Dentistry, St. Vincent Hospital, Catholic University of Korea, Ji-dong, Paldal-gu, Suwon 442-723, Republic of Korea

Kaufui V. Wong
Department of Mechanical and Aerospace Engineering, University of Miami, Coral Gables, FL 33146, USA

Aldo Hernandez
Department of Mechanical and Aerospace Engineering, University of Miami, Coral Gables, FL 33146, USA

V. Schirosi
Microelectronic Research, OPTEL InP Consortium Microelectronic Research Lab, c/o Cittadella della Ricerca, S.S. 7 Km 7.3, 72100 Brindisi, Italy

G. Del Re, L
Microelectronic Research, OPTEL InP Consortium Microelectronic Research Lab, c/o Cittadella della Ricerca, S.S. 7 Km 7.3, 72100 Brindisi, Italy

Ferrari, P. Caliandro
Microelectronic Research, OPTEL InP Consortium Microelectronic Research Lab, c/o Cittadella della Ricerca, S.S. 7 Km 7.3, 72100 Brindisi, Italy

L. Rizzi
Microelectronic Research, OPTEL InP Consortium Microelectronic Research Lab, c/o Cittadella della Ricerca, S.S. 7 Km 7.3, 72100 Brindisi, Italy

G. Melone
Microelectronic Research, OPTEL InP Consortium Microelectronic Research Lab, c/o Cittadella della Ricerca, S.S. 7 Km 7.3, 72100 Brindisi, Italy

Samuel Kenzari
Institut Jean Lamour, UMR 7198 CNRS-Université de Lorraine, F-54011 Nancy, France

David Bonina
Institut Jean Lamour, UMR 7198 CNRS-Université de Lorraine, F-54011 Nancy, France

Jean Marie Dubois
Institut Jean Lamour, UMR 7198 CNRS-Université de Lorraine, F-54011 Nancy, France

Vincent Fournée
Institut Jean Lamour, UMR 7198 CNRS-Université de Lorraine, F-54011 Nancy, France

Linan Zhu
Key Laboratory of Special Purpose Equipment and Advanced Processing Technology, Ministry of Education, Zhejiang University of Technology, Hangzhou 310014, China
School of Computer Science and Technology, Zhejiang University of Technology, Hangzhou 310023, China
College of Educational Science and Technology, Zhejiang University of Technology, Hangzhou 310023, China

Yanwei Zhao
Key Laboratory of Special Purpose Equipment and Advanced Processing Technology, Ministry of Education, Zhejiang University of Technology, Hangzhou 310014, China

Wanliang Wang
School of Computer Science and Technology, Zhejiang University of Technology, Hangzhou 310023, China

Rajeev Saha
Department of Mechanical Engineering, YMCA University of Science and Technology, Sector-6, Faridabad 121006, Haryana, Indi

Sandeep Grover
Department of Mechanical Engineering, YMCA University of Science and Technology, Sector-6, Faridabad 121006, Haryana, India

Alfred T. Sidambe
Bioengineering & Health Technologies Group, School of Clinical Dentistry, University of Sheffield, 19 Claremont Crescent, Sheffield S10 2TA, UK;

Hongyi Sun
Department of Systems Engineering and Engineering Management City University of Hong Kong China

Josef Hynek
University of Hradec Králové, Faculty of Informatics and Management Czech Republic

Václav Janeček
University of Hradec Králové, Faculty of Informatics and Management Czech Republic

Carlos Mireles
University Campus STeP Ri Slavka Krautzeka 83/A 51000 Rijeka, Croatia

Alfonso Noriega
University Campus STeP Ri Slavka Krautzeka 83/A 51000 Rijeka, Croatia

Gerardo Leyva
University Campus STeP Ri Slavka Krautzeka 83/A 51000 Rijeka, Croatia

F. Javier Ramírez
EXPAL Systems, Avenida Partenón 16, 28042 Madrid, Spain
School of Industrial Engineering, University of Castilla-La Mancha, Albacete, Spain

Rosario Domingo
School of Industrial Engineering, UNED, 28042 Madrid, Spain

Michael S. Packianather
School of Engineering, Cardiff University, Cardiff, UK

Miguel A. Sebastian
School of Industrial Engineering, UNED, 28042 Madrid, Spain

Preface

The text *Advanced Processing and Manufacturing Technologies* discusses the most important aspects necessary for understanding and development of processing and manufacturing of materials and systems. The latest developments in processing and manufacturing technologies are covered, including advanced composite manufacturing, rapid processing, joining, machining, and net shape forming technologies. First chapter focuses on assimilation patterns in the use of advanced manufacturing technologies in small and medium-sized enterprises. The purpose of second chapter is to provide a comprehensive viewpoint of issues related to successful AMT implementation and offer some directions to managers and investigators to make a company well-prepared to accept technology. The aim of third chapter to investigate the marginal fit of metal-free crowns made by three different computer-aided design/computer-aided manufacturing (CAD/CAM) systems. A review of additive manufacturing has been presented in fourth chapter. A novel manufacturing technology for RF MEMS devices on ceramic substrates has been focused in fifth chapter. Sixth chapter deals with complex metallic alloys as new materials for additive manufacturing. A bilayer manufacturing resource model with separation of cloud end and cloud manufacturing platform has been proposed in seventh chapter. The aim of eighth chapter is to assimilate the key enablers of e-manufacturing. The goal of ninth chapter is to summarize existing literature and report on the use of advanced manufacturing to fabricate titanium alloy implants. Tenth chapter deals with linking process technology and manufacturing performance under the framework of manufacturing strategy. Eleventh chapter focuses on specific issues of advanced manufacturing technology justification. Twelfth chapter discusses on flexible manufacturing system simulation using petri nets. In last chapter, an algorithm that allows the reduction of the total process times and cost in the manufacturing of axisymmetric components has been presented.

Chapter 1

ASSIMILATION PATTERNS IN THE USE OF ADVANCED MANUFACTURING TECHNOLOGIES IN SMES: EXPLORING THEIR EFFECTS ON PRODUCT INNOVATION PERFORMANCE

Sylvestre Uwizeyemungu[1], Placide Poba-Nzaou[2], Josée St-Pierre[1]

[1]Université du Québec à Trois-Rivières, Trois-Rivières, Canada
[2]Université du Québec à Montréal, Montréal, Canada

ABSTRACT

Manufacturing small and medium-sized enterprises (SMEs) are more and more adopting advanced manufacturing technologies (AMT) aimed at fostering product innovation process, improving product quality, streamlining the production process, and gaining productivity. In this study, we analyze the relationship between AMT proficiency levels in manufacturing SMEs and product innovation performance. Using data from 616 manufacturing SMEs, and considering a wide range of various AMT (20 different types of AMT grouped into 5 categories), we derived three AMT assimilation patterns through a cluster analysis procedure combining hierarchical and non-hierarchical clustering algorithms. The analysis of the relationship between AMT assimilation patterns and product innovation performance shows a rather unexpected picture: in spite of the existence of clearly distinct patterns of AMT assimilation, we find no significant relationship between any pattern and product innovation performance. Instead, we find the organizational and environmental context of SMEs to be more determinant for product innovation performance than any of the AMT assimilation patterns. From a practical point of view, this study indicates that manufacturing SMEs managers interested in fostering their innovation capabilities through AMT assimilation need to be aware of the contingency effects of their organizational size, age, and sector of activity.

INTRODUCTION

The need to meet the requirements of demanding customers and keep up with tough competition puts pressure on manufacturing small and medium-sized enterprises (SMEs) to offer high-quality and innovative products while being efficient in their production processes. The intensity of this pressure is such that SMEs have to simultaneously excel in various areas, without making trade-offs. The implementation of information technologies (IT), particularly advanced manufacturing technologies (AMT), has been seen as an important step forwards for SMEs in their quest to achieve a wide range of organizational and technological benefits such as innovation and productivity that help them deal with the market pressure they experience.

However, it has been argued that the effective implementation of quality improvement approaches or tools (including AMT) in SMEs is hindered by these firms' limited resources and knowledge. Hence the need to account for the specificities of SMEs when analyzing the contribution of AMT to organizational performance. One cannot simply assume that the types of AMT that are advantageous in the context of large enterprises (LE), the underlying models of adoption and exploitation, and the results achieved would be necessarily transposable as such to SMEs. Differences between LEs and SMEs with regards to resource constraints, organizational contingencies, as well as entrepreneurial and strategic choices are likely to play a determinant role when it comes to antecedents and outcomes of AMT implementations.

Manufacturing SMEs, given their constraints, particularly in terms of resources and expertise, cannot always invest in a wide range of AMT. They need to decide carefully among the multiple AMT available to which they will dedicate their limited resources. Which AMT does it matter for manufacturing SMEs to highly assimilate? More precisely, for which types of AMT do SMEs need to develop high levels of proficiency in order to reach higher performance levels in terms of product innovation? Are there any particular AMT assimilation patterns that could be associated with product innovation performance?

We explore these questions through an analysis of data collected from a sample of 616 manufacturing SMEs. We derived three AMT assimilation patterns (three clusters) through a cluster analysis procedure combining hierarchical and non-hierarchical clustering algorithms. In spite of clearly distinct patterns of AMT assimilation, we find no significant relationship between any patterns and related product innovation performance. Instead, we find that the organizational and environmental contexts of SMEs are more determinant for product innovation performance than any of the AMT categories.

The remainder of this paper is organized as follows: in section 2 we present the research's theoretical and empirical background leading to our research model. We set forth our research methodology in section 3. We reveal the results of our analysis in section 4 and we analyze and discuss them in section 5. It is also in this last section that we discuss the theoretical and practical implications of our results, the research limitations, and the research avenues.

THEORETICAL AND EMPIRICAL BACKGROUND

Advanced Manufacturing Technologies: Definition and Classification

The various technologies grouped under the label AMT and broadly defined by OECD as "computer-controlled or micro-electronics-based equipment used in the design, manufacture or handling of a product" are diversely classified in literature. For instance distinguished (1) stand-alone, (2) intermediate, and (3) integrated technologies whereas proposed a three-category typology of advanced manufacturing systems (AMS): (1) product design technologies (AMS for innovation), (2) process technologies (AMS for flexibility), and (3) logistics/planning applications (AMS for integration). This last categorization is particularly interesting as it implies the objectives pursued by adopting organizations. In this study, we adopt this last classification with a slight variation. We split the last category of logistics/planning applications (IT for integration) into two parts, by removing transactional and logistic applications and grouping them under the label of AMT for logistic and monitoring, while keeping the remaining applications under the label of AMT for integration. In Table 1 we present the four categories of AMT that are used in the present study.

Table 1: Classification of Advanced Manufacturing Technologies (AMT)

Product Design Technologies (AMT for Innovation):
Computer-aided design and manufacturing (CAD/CAM)
Computer-aided design (CAD)
Computer-aided drawing
Computer-aided manufacturing (CAM)

Process Technologies (AMT for Flexibility):
Flexible manufacturing systems (FMS)
Programmable logic controller (PLC)
Automated handling of materials
Computer numerically controlled machines (CNC)
Robotized operations
Transactional and Logistic Applications (AMT for logistic and monitoring):
Bar codes
Production inspection & control
Computer-based production scheduling
Computer-aided maintenance
Quality control system
Computer-based inventory management
Communication and Integration Applications Systems (AMT for Integration):
Enterprise resource planning (ERP)
Material requirement planning (MRP-I)
Manufacturing resource planning (MRP-II)
Electronic data interchange (EDI)
LAN for MRP-II/Plant/Intranet

AMT in SMEs

AMT Adoption Challenges in SMEs

SMEs aiming at adopting AMT are confronted with challenges related to their internal and external environments. With regards to internal environment, the lack of resources impedes even the appropriate assessment of the benefits SMEs could reap from the implementation of AMT. Indeed, the selection of a number of AMT to be implemented in a firm is part of a process towards a targeted model for an organizational design that managers think is compatible with external market requirements. AMT are thus resources meant to achieve specific manufacturing capabilities (internal requirements) that would allow the adopting firms to achieve improved operational and strategic results. This difficulty is part of a broader set of problems faced by SMEs in their quest to adopt and exploit AMT. In general, most SMEs lack what refer to as "assets for AMT", that is the resources necessary for properly developing and using AMT. In other words, SMEs are not only deterred by the extensive capital investment necessary to adopt and implement AMT, but they also lack technical and manufacturing infrastructure to support these technologies. In this sense, the notion of "assets for AMT" is close to the concept of absorptive capacity.

Beyond resources, there are also other internal factors that play a determinant role in the adoption of AMT in SMEs. For example, the characteristics of the entrepreneur (education level, experience in industry) and the entrepreneurial strategic orientation (development on new markets, adoption on new technologies) have been found to influence the nature of AMT adopted in SMEs. Interestingly, it also appears that the nature of operational performance achieved with the initial AMT adopted by SMEs has a crucial role with regard to additional investments in AMT: improvements in quality and flexibility induce firms to adopt further AMT, while achievements in terms of low-cost, innovation, and delivery capability do not lead to further AMT adoption n (Spanos & Voudouris, 2009).

The levels of adoption of AMT are also influenced by the external environment of SMEs (e.g. customers and vendors). SMEs exposed to some types of AMT through their relationship with their customers or suppliers will likely tend to adopt the same technologies. The type of production (sector of activity) and the commercial dependency have also been identified by as other environmental factors that can influence SMEs with respect to AMT adoption.

State of AMT Adoption in SMEs

It is not easy to get from the literature on the subject a more or less complete picture of the state of AMT adoption in SMEs. Studies do not use the same list of technologies, which makes it difficult to compare their results. However, in spite of this limitation, a consistent conclusion stems from previous studies: SMEs, generally undercapitalized and resource-constrained, tend to adopt only a few types of AMT. identified three main types of AMT common in SMEs: LAN (Local Area Network), CAD, and CAM. Automated storage, robotics, and WAN (Wide Area Network) were found to be the less commonly diffused in sampled SMEs. According to the authors, the adoption of LAN was a response to a need by the SMEs to integrate all their functions under a computer network in order to achieve better performance outcomes. As for the adoption of CAD and CAM, it is consistent with the nature of the job shop processes in SMEs, characterized by a high degree of customization. CAD and CAM would allow SMEs to meet the design challenges stemming from the differentiation requirements of their markets. In another study, identified computer-based inventory management, computer-aided drawing (CAD), and equipment controlled by programmable automata as the most prevalent AMT of the study's sampled SMEs. In the same study, flexible manufacturing systems (FMS), and MRP-II were the least common. The results of this study show that no category of AMT among the four categories we defined earlier (cf. Table 1) is predominantly adopted by SMEs. They rather reflect a "pick and choose" strategy amongst different categories of AMT.

AMT and Firm Performance

Prior studies examining the performance effects of AMT adoption and usage have produced conflicting results. While some reported positive results, others reported negative results. For example, when comparing two samples of firms from Sweden and Singapore, found that AMT investments were correlated with firm performance in Sweden, but not in Singapore. In a longitudinal study based on data collected from 308 companies over 22 years, did not find a statistically significant relationship, direct or indirect, between AMT and productivity. Overall, failures in AMT projects are more common than successes, and the situation appears to be more dramatic in SMEs.

The analysis of the research results that report positive impacts of AMT shows that the AMT – performance relationship is complex. Some studies have established that the AMT-performance relationship is either mediated or moderated by other factors such as: operation improvement practices, quality management practices, workers' empowerment, etc. The requirement of complementary resources is another facet of the complexity of AMT – performance relationship. In accordance with the resource-based theory, complementary resources or investments are generally necessary and sometimes indispensable for organizations to take advantage of their AMT investments. Investing in training and mentoring, as well as in developing close buyer-supplier relationships improves the likelihood of better AMT performances. It would also be beneficial to invest in organizational structure adaptation: the low levels of productivity gained from AMT even several years after their implementation in Indian manufacturing firms are attributed to the incompatibility of new technologies with the organizational structure that has not evolved (it was and remained mechanistic). In other studies (Lewis & Boyer, 2002;) it has been found that the business value of AMT investments is conditional to the implementation efforts and to the deployment of the appropriate implementation strategies.

The above considerations highlight the importance of appropriateness of AMT with regards to the organizational context; which leads to the concept of "fit" or "alignment". In accordance with the contingency theory perspective, the effects of AMT on organizational performance would stem from their fit or alignment with other organizational dimensions. For example, it has been found that the "fit" (conceptualized as profile deviation) between process environment and AMT leads to superior performance, and the "mismatch", which is the deviation from ideal profiles, has a negative impact on manufacturing performance. In other studies, the alignment between strategies and different types of AMT, and the alignment of SMEs' network, product and market development with their levels of AMT integration and assimilation were associated with better performance.

The availability of several types of AMT means that companies can combine them in various ways. Hence the importance of identifying AMT investment patterns and their respective impact on performance outcomes. Identified three groups corresponding to different patterns of AMT investment behavior (groups labelled traditionalists, designers, and investors). However, they were unable to relate different performance levels to the identified groups. Later on, in other studies, differences in performance results were related to different AMT investment patterns.

Research Model

In this study, we analyze the patterns of AMT assimilation in SMEs, as well as their corresponding performance in terms of product innovation. Our research model (Figure 1) is based on the assumption that firms choose among diverse available types of AMT a set of technologies more or less assorted. They then implement those technologies with great or less success in terms of assimilation (or proficiency in use). We postulate that it is the level of assimilation of the various adopted AMT that determines the level of product innovation achieved.

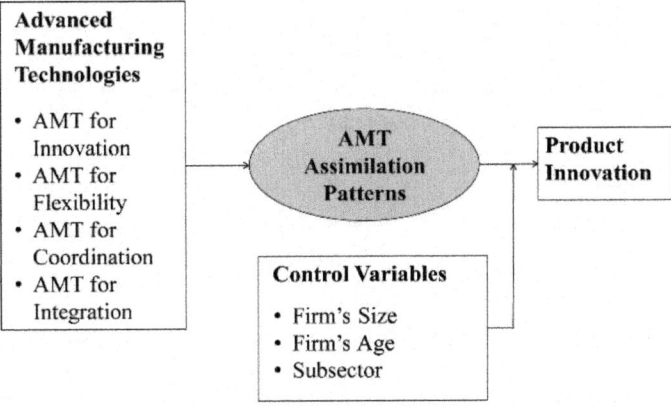

Figure 1: Research Model

As the implementation of AMT occur in a context that can moderate their effects on organizational performance, our research model takes into account some contextual variables, namely the firm's size and age, and its subsector of activity. Previous studies have established that the firm's size, the firm's age, and the nature of activity have an impact on the adoption of technological innovations and on the latter's effects on performance outcomes.

RESEARCH METHOD

Data Source

For the purpose of this study we used data previously collected from manufacturing SMEs by a university research center. In cooperation with an important business association, this center has created a database that contains information on a wide range of business practices and on various aspects of these SMEs' organizational performance. SME owner-managers and key employees filled out a questionnaire and provided their firm's financial statements for the last three years. On the basis of the information supplied, a comparative diagnosis (benchmarking) of their firm is provided to owner-managers, assessing the firm's overall situation in terms of performance and vulnerability. These diagnoses link the SMEs' results to their organizational resources and business practices (including advanced manufacturing technologies), and formulate recommendations as to what actions should be undertaken in short and medium terms to improve performance and/or reduce vulnerability.

Sample

For this study's purposes we retrieved from the database data on 673 manufacturing firms, and we retained 616 observations after removing cases with under 10 employees (42 cases) or over 250 employees (15 cases). The upper limit of 250 employees corresponds to the OECD definition of SMEs based on size. Manufacturing firms under 10 employees were excluded as being too small with limited likelihood of interest in AMT. The analysis of missing values in our data set (using the SPSS Missing Values Analysis - MVA - algorithm) shows that missing values represent 840 of the 13,944 values (cases x variables), that is 5.68%. This rate falls in the range of 5 to 15%, which implies that the handling of missing values requires sophisticated methods. Considering this result, we decided to estimate missing values even if the Little's MCAR test succeeded (p-value of 0.463 > 0.05). We estimated missing values using multiple imputation technique as recommended in such a situation.

The mean number of employees of the 616 manufacturing SMEs included in our sample is 54 employees (standard deviation: 45), and the median is 40 employees. The mean age of SMEs in our sample is 38.6 years (s.d.: 23.3), with a median of 33 years.

The manufacturing SMEs studied are spread over 20 sub-sectors. The most represented sub-sectors include metal products (37.7%), rubber and plastic industry (12.7%), wood industry (9.1%), electrical and electronic products (6.5%), machinery (6.2%), and food products and beverages (5.8%). When grouped according to the OECD's four-category classification of industrial sectors based on the technological intensity SMEs in our sample are divided into the following proportions: there are 166 (26.9%) low-tech SMEs, 351 (57.0%) medium to low-tech SMEs, 99 (16.1%) medium to high-tech SMEs. There are no SMEs in high-tech category. Finally, it is worth mentioning that although our sample consists mainly of Canadian firms (433 – 70.3%), it also includes French (172 – 27.9%) and Mexican companies (12 – 1.8%).

Measurement

Measures used for all of our variables (clustering variables, organizational performance variables, and control variables) are presented in Table 2.

Table 2: Variables Measurement

Category	Variable	Measure	References
Clustering Variables	Assimilation levels of 20 different AMT adopted	Proficiency in use of each AMT, on a scale of 1 (low) to 5 (high). Whenever a given AMT is not present, the proficiency score is 0.	Brandyberry et al. (1999); Raymond and Croteau (2009).
Organizational Performance Variable	Innovation	Average percentage of sales attributed to new or modified products over the last two financial years.	Becheikh et al. (2006); Garcia et al. (2002).
Control Variables	• Firm size	Average number of employees over the two last periods.	-
	• Firm age	Years of existence from the year of creation up to 2014 (2014-year of creation).	-
	• Sub-sector	OECD classification of industrial activities based on technological intensity: low-tech (1), medium to low-tech (2), medium to high-tech (3), and high-tech (4).	OECD (2005).

Cluster Analysis

As previously stated, we used, as clustering variables, the levels of assimilation of various advanced manufacturing technologies (AMT) adopted by SMEs. We preferred levels of assimilation measured on an ordinal scale over a dichotomous measure of AMT adoption (0/1) as we anticipated, following previous studies, that assimilation matters more than mere adoption.

Following 's recommendation, we combined two clustering methods: we first used the hierarchical clustering algorithm (agglomerative method) to decide upon the "optimal" number of clusters and determine the clusters centroids, and we then used these results as initial seeds for a non-hierarchical algorithm. To determine the "optimal" number of clusters, we applied the agglomerative hierarchical clustering algorithm (Ward's method) on our complete sample, and based upon the analysis on the dendrogram produced, we identified four plausible solutions, a 2-cluster solution, a 3-cluster solution, a 4-cluster solution, and an 8-cluster solution.

For determining among those solutions the one that would be stable, we randomly selected a first sub-sample of approximately 50% of our sample, and then a second sub-sample composed of approximately 30% of the full sample (using SPSS random selection procedure), and we applied the same clustering algorithm. The analysis of the dendrograms produced with the two sub-samples data indicated that the 3-cluster solution was the most stable.

In the second phase of our cluster analysis, we applied the K-Means clustering algorithm, using K=3, and the mean values produced in the 3-solution cluster as initial seeds for the algorithm. Although the iteration number was tentatively set at 100, the clustering results were convergent after 11 iterations.

RESULTS

State of AMT Use in SMEs

The four most prevalent AMT in the sampled SMEs are, in descending order: production inspection and control system (67.2%), computer-based inventory management (66.6), computer-aided drawing (64.1%), and quality control system (61.9%). It is worth noting that three out of four AMT that are present in more than 60% of SMEs are in the category of AMT for logistic and monitoring. The least prevalent AMT are MRP-II (17.9%), FMS (18.7), automated handling of materials (20.0%), and computer-based maintenance management (20.1%). AMT for integration registered the lowest rates of adoption in the sampled SMEs, closely followed by AMT for flexibility. The components of AMT for integration category are adopted by less than 30% of SMEs, except for ERP (41.4%).

Results of Cluster Analysis

We present in Table 3 the final results of the earlier described clustering process. Our 616 SMEs are divided into three clusters, 170 (27.6%) in cluster I, 162 (26.3%) in cluster II, and 284 (46.1%) in cluster III. Overall, SMEs in cluster I display the strongest AMT assimilation: they come first in three categories out of four, and second in the remaining category (AMT for logistic and monitoring). The lowest AMT assimilation levels are displayed by SMEs in cluster III: they come in the second position only in the category of AMT for innovation, and they come in the last position in the other three categories. SMEs in the cluster II fall globally in the middle position: they are, however, the strongest in the category of AMT for logistic and monitoring, and the weakest in the category of AMT for Innovation, and they come in the second position for the remaining categories.

Table 3: AMT Assimilation Patterns Resulting from the Cluster Analysis

Clustering Variables	Clusters			Anova
	I (n = 170) Mean	II (n = 162) Mean	III (n = 284) Mean	F
AMT for Innovation	High	Low	Medium	
• Computer-Aided Design and Manufacturing (CAD/CAM)	2.49$_a$	0.15$_c$	0.36$_b$	194.7***
• Computer-Aided Design (CAD)	3.53$_a$	0.54$_c$	1.38$_b$	140.6***
• Computer-Aided Drawing	4.07$_a$	1.47$_c$	2.18$_b$	97.9***
• Computer-Aided Manufacturing (CAM)	3.39$_a$	0.41$_b$	0.58$_b$	274.9***
AMT for Flexibility	High	Medium	Low	
• Flexible Manufacturing Systems (FMS)	0.97$_a$	0.57$_b$	0.27$_c$	17.4***
• Programmable Logic Controller (PLC)	2.17$_a$	2.78$_b$	0.59$_c$	92.1***
• Automated Handling of Materials	0.98$_a$	1.35$_a$	0.17$_b$	39.5***
• Computer Numerical Control (CNC)	3.56$_a$	1.38$_b$	0.85$_c$	138.3***
• Robotics	1.49$_a$	1.47$_a$	0.30$_b$	45.7***
AMT for Logistic and Monitoring	Medium	High	Low	
• Bar Codes	1.56$_a$	1.60$_a$	0.40$_b$	40.5***
• Production Inspection & Control	2.74$_b$	3.40$_a$	1.54$_c$	70.4***
• Production Scheduling	2.17$_a$	1.83$_a$	0.56$_b$	63.1***
• Computer-Aided Maintenance	0.97$_a$	0.56$_b$	0.15$_c$	32.3***
• Quality Control System	2.89$_b$	3.33$_a$	1.18$_c$	105.1***
• Inventory Management System	2.76$_b$	3.23$_a$	1.29$_c$	87.2***
AMT for Integration	High	Medium	Low	
• ERP	1.12$_a$	1.26$_a$	0.53$_b$	20.3***
• MRP-I	1.45$_a$	1.10$_a$	0.34$_b$	35.8***
• MRP-II	0.85$_a$	0.51$_a$	0.12$_b$	27.2***
• EDI	1.30$_a$	1.48$_a$	0.33$_b$	37.6***
• LAN for MRP-II/Plant/ Intranet	1.40$_a$	0.87$_b$	0.26$_c$	42.2***

***: $p < 0.001$ (two-tailed tests)
a, b, c: Within rows, different subscripts indicate significant ($p < 0.05$) pair-wise differences between means on Tamhane's T2 (post hoc) test

The pair-wise differences between means on Tamhane's T2 post-hoc test (cf. subscripts following the means in Table 3) allow us to identify significant differences between the means of different clusters. Being in the middle position, cluster II offers a good starting point for a pair-wise comparison. The scrutiny of the results of Tamhane's T2 post-hoc test shows that the big difference between cluster II and cluster I lies in their respective levels of assimilation of mainly AMT for innovation, and secondarily AMT for flexibility, and AMT for logistic and monitoring. Their differences in terms of assimilation of AMT for integration are quite minimal: in this case, the sole component that is statistically distinctive when clusters I and II are compared is LAN for MRP-II/Plant/Intranet.

The comparison between cluster II and III shows a quite different situation: the two clusters are significantly different on almost all AMT. Actually, there is only one AMT on a total of 20 on which the two groups are not significantly distant one from the other: they are statistically similar on their assimilation levels of only computer-aided manufacturing (CAM). Not surprisingly, cluster I and cluster III are far apart on the whole range of AMT.

In Tables 4 and 5, we present the breakdown of control and organizational performance variables by AMT assimilation patterns. Table 5 shows the results for the sub-sector, and Table 4 shows the results for the rest of the control and organizational performance variables. These results allow us to analyze differences between the three clusters with regard to variables "theoretically related to the clusters, but not used in defining clusters" (Ketchen Jr. & Shook, 1996, p. 447). Here again, the results of the Tamhane's T2 post-hoc test help us establish pair-wise differences.

Table 4: Breakdown of Control and Organizational Performance Variables by AMT Assimilation Patterns

	Cluster				
Variables	I	II	III	Total	Anova
	$(n = 170)$	$(n = 162)$	$(n = 284)$	$(n = 616)$	
	Mean	Mean	Mean	Mean	F
Control Variables					
• Firm's size	70.93_a	60.05_a	39.91_b	53.76	29.9***
• Firm's age	41.87_a	$40.03_{a,b}$	35.89_b	38.63	4.0*
Performance Variable					
• Innovation	0.12	0.12	0.10	0.11	1.7

*: $p < 0.05$; ***: $p < 0.001$ (2-tailed tests)

a, b, c: Within rows, different subscripts indicate significant ($p < 0.05$) pair-wise differences between means on Tamhane's T2 (post hoc) test.

Table 5: Breakdown of Sub-Sectors by AMT Assimilation Patterns

Sub-Sector	Cluster								χ^2
	I		II		III		Total		
	f	%	f	%	f	%	f	%	
• Low-Tech	24	3.9	55	8.9	87	14.1	166	26.9	
• Medium to Low-Tech	127	20.6	89	14.5	135	21.9	351	57.0	39.58***
• Medium to High-Tech	19	3.1	18	2.9	62	10.1	99	16.1	
Total	170	27.6	162	26.3	284	46.1	616	100	

***: $p < 0.001$

With regard to the firm's size, clusters I and II are statistically comparable, even though SMEs in cluster I are relatively larger than SMEs in cluster II (mean size of respectively 71 and 60 employees). SMEs in cluster III are the smallest (40 employees on average), and significantly different from SMEs in the other two clusters. Regarding the age of firms, clusters I and II, with the mean age of 41.8 and 40 years respectively, are comparable and above the mean age of the entire sample (38.6 years). SMEs in cluster III, with the mean age of 35.6 years, are younger than SMEs in clusters I and II. The difference between clusters with respect to age is statistically significant only between cluster I and III.

The breakdown of sub-sectors by AMT assimilation patterns presented in Table 5 shows that the different sub-sectors are not randomly distributed into the three clusters (global chi-square of 39.58, $p < 0.001$). To further clarify this conclusion, we present in Table 6 what expected percentages of each sub-sector in the three clusters would be if the distribution were random. These expected percentages take into account the actual distributions of SMEs in different sub-sectors and in different clusters. The comparison of percentages in Tables 5 and 6 shows that low-tech SMEs are under-represented in the cluster I (and inversely over-represented in the clusters II and III) while medium to low-tech SMEs are over-represented in the cluster I (and inversely under-represented in the clusters II and III). Medium to high-tech firms are over-represented in the cluster III (and inversely under-represented in the clusters I and II).

Table 6: Expected Distribution (%) of SMEs Pertaining to Different Subsectors in Different Clusters under the Hypothesis of Random Distribution

	Cluster (%)		
Sub-Sector (%)	I (27.6)	II (26.3)	III (46.1)
• Low-Tech (26.9)	7.4	7.1	12.4
• Medium to Low-Tech (57.0)	15.7	15.0	26.3
• Medium to High-Tech (16.1)	4.4	4.2	7.4

With regard to product innovation performance, the ANOVA results (Table 4) show that there are no significant differences between clusters. These results are further discussed in section 5.

DISCUSSION OF THE RESULTS

Analysis of the Results

Do the results indicate that advanced manufacturing technologies (AMT) play a determinant role in fostering product innovation performance? The cluster analysis clearly shows that there are three distinct AMT assimilation patterns amongst the sampled SMEs, but the relationship between these patterns and product innovation performance seems to be elusive.

As apparent in Table 4, SMEs in cluster III tend to be smaller (mean size of 40 employees) than SMEs in other clusters (60 employees in cluster I, and 71 employees in cluster II). However, the difference between clusters I and II in terms of size is not statistically significant. As small firms experience a greater impact from their adoption of innovations (including technologies), they would need lower proficiency levels to achieve good results in terms of product innovation. Their small size allows them to be more reactive to changes in their environment than bigger enterprises would be. In addition, the proximity of smaller firms to their market is another factor conducive to their being innovative and effective. This situation compensates for their weak mastery of AMT.

Firms in cluster I are the oldest (mean age of 41.9 years), although the age variance between them and firms in cluster II (40 years) is not statistically significant. The age variance between these two clusters (I and II) and cluster III (35.9) is significant, and this difference may explain why firms in cluster III perform relatively well: they achieve comparable results with firms in the other two clusters in spite of their being less proficient in terms of AMT. As innovation propensity is higher in younger firms than in older ones, younger firms in cluster III may achieve good innovation results even with their lower levels of proficiency with AMT. The results of this study seem to confirm the observation that the forces of organizational inertia that increases as firms get older prevail over the benefits of organizational learning gained with time. However, our conclusion in this respect is to be taken with caution given that the "benefits of organizational learning" are limited to product innovation in our study.

Another possible factor that explains the results as far as product innovation is concerned would be related to environmental uncertainty. Organizations operating in a highly uncertain environment are likely to be compelled to innovate at a higher rate than organizations in a less uncertain environment. We do not have a measure of environmental uncertainty in our research variables, but the industrial sector (measured through technological intensity) can be used as a proxy for such a measure. The over-representation of SMEs in the category of medium to high-tech sub-sector (the highest technology-intensive sector in our sample) in the cluster III would explain the relatively good performance of this group on product innovation performances. However, this explanation is weakened by the observation that the predominance of low-tech SMEs in cluster II does not prevent them from achieving comparable results as the other groups in terms of product innovation.

Instead of considering the overall AMT assimilation levels, one could try to analyze the assimilation levels by AMT category, and see if some subtle effects of each category on product innovation performance would be found. As SMEs in cluster I display the highest levels of assimilation of AMT for innovation, one would expect them to perform better on innovation performance, at least when compared to SMEs in cluster II which display the lowest assimilation levels in the same category. The actual results show that the expected relationships do not necessarily materialize. The fact that SMEs with the highest assimilation levels in AMT for innovation (cluster I) do not show outstanding innovation performance, as it was expected, could be explained by the negative moderating effect of AMT for integration, category in which the same SMEs display the highest assimilation levels. Indeed, it has been noted that high levels of information technology (IT) integration tend to negatively affect the market-oriented flexibility of the production process and to disable SMEs' innovation capability effects on productivity.

Implications, Limitations, and Research Avenues

Theoretical and practical implications can be drawn from this study. The first theoretical contribution of this study is to make relative the cause and effect relationship between advanced manufacturing technologies (AMT) and product innovation performance, at least in the context of small and medium-sized enterprises (SMEs). Notably, the results from this study show that for product innovation in SMEs the organizational and environmental context is more determinant than the actual levels of AMT assimilation.

As our second theoretical contribution, we have used the measures of AMT assimilation as independent variables instead of the more usual IT adoption. When using AMT adoption as independent variables, firms are split into two categories, the adopters and non-adopters. The assimilation measure allows us to fine-tune the analysis, as one passes from a 0-1 scale to a 0-5 scale, allowing for the application of a form of gradation among firms that have adopted the technology but display varying degrees of proficiency in the use of that technology.

The third contribution of this study stems from the spectrum of the AMT covered. Twenty different AMT grouped into four categories are included in this study. Taking into consideration multiple AMT allows for the considerations of the potential synergistic and conflicting effects of different technologies. It also allows for the analysis of the effectiveness of various resource configurations (or resource patterns). As AMT are available from vendors, any firm, provided that it has access to necessary resources, can choose any among them. Therefore, the AMT-related competitive advantage does not reside into the technologies themselves, but in their ingenious combination; a combination that would be hard for competitors to imitate. This is consistent with the resource-based view (RBV), a theoretical framework that is widely used to explain the impact of IT on organizational performance .

From a practical point of view, this study elucidates the fact that manufacturing SMEs managers interested in fostering their innovation capabilities through AMT assimilation need to be aware of the determinant role of their organizational and environmental contexts. When deciding for which types of AMT they will allocate their limited resources, they should take into account the contingency effects of their organizational size, age, and sector of activity. For instance, owner-managers and manufacturing plants managers in SMEs, in particular older SMEs, should be aware of the damaging effects of organizational inertia phenomena. This study suggests that this phenomena could, in the long run, lead to the erosion of benefits due to the implementation and exploitation of AMT. SMEs' managers will have to find ways to thwart such erosion. For AMT vendors and consultants, this study signals a need for them to develop a differentiating scale of offers for SMEs with regard to the various organizational and environmental contexts. The first limitation of this study relates to the AMT assimilation measures that are based on self-reported ratings from operations and manufacturing managers. These subjective measures may lead to bias given that different individuals may diversely judge the levels of proficiency. Another limitation is not having considered whether SMEs are subcontractors or not, while this status may influence the production systems and the type of innovation adopted. In future research, the mediating role of this status may be taken into consideration.

As organizational performance indicator, we have used product innovation. Even though product innovation is a good indicator of manufacturing SMEs' performance, one could choose other performance indicators such as product and process quality, productivity, profitability, sales revenue, growth, etc. AMT effects may vary according to the business performance indicators one considers. This consideration applies also to the control variables. For instance, it would have been interesting to have data on environment uncertainty, and analyze its effects on the relationship between AMT assimilation patterns and organizational performance. In the same way, it would be interesting to analyze the mediating role of networks in which SMEs are involved. As a research avenue, future research could analyze the AMT assimilation patterns on other performance measures, and include other control variables. Additionally, future research could focus on the efforts to identify the effective combinations of different AMT in SMEs. The results of this kind of research will allow SMEs to wisely spend their limited resources on AMT that best provide for the achieving of the desired performances.

REFERENCES

1.   Acock, A. C. (2005). Working With Missing Values. *Journal of Marriage and Family, 67*(4), 1012–1028.

2.   Acuña, E., & Rodriguez, C. (2004). The Treatment of Missing Values and its Effect in the Classifier Accuracy. In D. Banks, L. House, F. R. McMorris, P. Arabie & W. Gaul (Eds.), *Classification, Clustering and Data Mining Applications* (pp. 639-648). Berlin-Heidelberg: Springer-Verlag.

3.   Balasubramanian, N., & Lee, J. (2008). Firm Age and Innovation. *Industrial and Corporate Change, 17*(5), 1019-1047.

4.   Balijepally, V., Mangalaraj, G., & Iyengar, K. (2011). Are We Wielding this Hammer Correctly? A Reflective Review of the Application of Cluster Analysis in Information Systems Research. *Journal of the Association for Information Systems, 12*(5), 375-413.

5.   Becheikh, N., Landry, R., & Amara, N. (2006). Lessons from Innovation Empirical Studies in the Manufacturing Sector: A Systematic Review of the Literature from 1993-2003. *Technovation, 26*(5/6), 644-664.

6. Bhatt, G. D., & Grover, V. (2005). Types of Information Technology Capabilities and their Role in Competitive Advantage: An Empirical Study. *Journal of Management Information Systems, 22*(2), 253-277.

7. Birdi, K., Clegg, C., Patterson, M., Robinson, A., Stride, C. B., Wall, T. D., & Wood, S. J. (2008). The Impact of Human Resource and Operational Management Practices on Company Productivity: A Longitudinal Study. *Personnel Psychology, 61*(3), 467-501.

8. Brandyberry, A., Rai, A., & White, G. P. (1999). Intermediate Performance Impacts of Advanced Manufacturing Technology Systems: An Empirical Investigation. *Decision Sciences, 30*(4), 993-1020.

9. Bülbül, H., Ömürbek, N., Paksoy, T., & Bektas, T. (2013). An Empirical Investigation of Advanced Manufacturing Technology Investment Patterns: Evidence from a Developing Country. *Journal of Engineering and Technology Management, 30*(2), 136-156.

10. Cardoso, R. D. R., Pinheiro de Lima, E., & Gouvea da Costa, S. E. (2012). Identifying Organizational Requirements for the Implementation of Advanced Manufacturing Technologies (AMT). *Journal of Manufacturing Systems, 31*(3), 367-378.

11. Choe, J.-M. (2004). Impact of Management Accounting Information and AMT on Organizational Performance.*Journal of Information Technology, 19*(3), 203-214.

12. Chung, W., & Swink, M. (2009). Patterns of Advanced Manufacturing Technology Utilization and Manufacturing Capabilities. *Production and Operations Management, 18*(5), 533-545.

13. Dangayach, G. S., & Deshmukh, S. G. (2005). Advanced manufacturing technology implementation: Evidence from Indian small and medium enterprises (SMEs). *Journal of Manufacturing Technology Management, 16*(5/6), 483-496.

14. Das, A., & Narasimhan, R. (2001). Process-Technology Fit and its Implications for Manufacturing Performance.*Journal of Operations Management, 19*(5), 521-540.

15. Desai, D. A. (2008). Cost of Quality in Small- and Medium-Sized Enterprises: Case of an Indian Engineering Company. *Production Planning & Control, 19*(1), 25-34.

16. Diaz, M. S., Machuca, J. A. D., & Alvarez-Gil, M. J. (2003). A view of Developing Patterns of Investment in AMT through Empirical Taxonomies: New Evidence. *Journal of Operations Management, 21*(5), 577-606.

17. Franquesa, J., & Brandyberry, A. (2009). Organizational Slack and Information Technology Innovation Adoption in SMEs. *International Journal of E-Business Research, 5*(1), 25-48.

18. Fulton, M., & Hon, B. (2010). Managing Advanced Manufacturing Technology (AMT) Implementation in Manufacturing SMEs. *International Journal of Productivity and Performance Management, 59*(4), 351-371.

19. Garcia, R., & Calantone, R. (2002). A Critical Look at Technological Innovation Typology and Innovativeness Terminology: A Literature Review. *Journal of Product Innovation Management, 19*(2), 110-132.

20. Ghani, K. A., Jayabalan, V., & Sugumar, M. (2002). Impact of Advanced Manufacturing Technology on Organizational Structure. *Journal of High Technology Management Research, 13*(2), 157-175.

21. Gresov, C., & Drazin, R. (1997). Equifinality: Functional Equivalence in Organization Design. *Academy of Management Journal, 22*(2), 403-428.

22. Gupta, A., & Whitehouse, F. R. (2001). Firms Using Advanced Manufacturing Technology Management: An Empirical Analysis Based on Size. *Integrated Manufacturing Systems, 12*(5), 346-350.

23. Huergo, E., & Jaumandreu, J. (2004). How Does Probability of Innovation Change with Firm Age? *Small Business Economics, 22*(3/4), 193-207.

24. Ketchen Jr., D. J., & Shook, C. L. (1996). The Application of Cluster Analysis in Strategic Management Research: An Analysis and Critique. *Strategic Management Journal, 17*(6), 441-458.

25. Koc, T., & Bozdag, E. (2009). The Impact of AMT Practices on Firm Performance in Manufacturing SMEs.*Robotics and Computer-Integrated Manufacturing, 25*(2), 303–313.

26. Kotha, S., & Swamidass, P. M. (2000). Strategy, Advanced Manufacturing Technology and Performance: Empirical Evidence from U.S. Manufacturing Firms. *Journal of Operations Management, 18*(3), 257-277.

27. Laforet, S. (2013). Organizational Innovation Outcomes in SMEs: Effects of Age, Size, and Sector. *Journal of World Business, 48*(4), 490–502.

28. Lagacé, D., & Bourgault, M. (2003). Linking Manufacturing Improvement Programs to the Competitive Priorities of Canadian SMEs. *Technovation, 23*(8), 705-715.

29. Laosirihongthong, T., & Himanghsu, P. (2004). Competitive Manufacturing Strategy: An Application of Quality Management Practices to Advanced

Manufacturing Technology Implementation. *International Journal of Business Performance Management, 6*(3,4), 262-286.

30. Lefebvre, L. A., Lefebvre, E., & Harvey, J. (1996). Intangible Assets as Determinants of Advanced Manufacturing Technology Adoption in SME's: Toward an Evolutionary Model. *IEEE Transactions on Engineering Management, 43*(3), 307-322.

31. Lewis, M. W., & Boyer, K. K. (2002). Factors Impacting AMT Implementation: An Integrative and Controlled Study. *Journal of Engineering and Technology Management, 19*(2), 111-130.

32. OECD. (2005). *Oslo Manual: Guidelines for Collecting and Interpreting Innovation Data* (OECD Ed. 3 ed.). Paris: OECD.

33. Patterson, M. G., West, M. A., & Wall, T. D. (2004). Integrated Manufacturing, Empowerment, and Company Performance. *Journal of Organizational Behavior, 25*(5), 641-665.

34. Payne, G. T. (2006). Examining Configurations and Firm Performance in a Suboptimal Equifinality Context.*Organization Science, 17*(6), 756-771.

35. Potthoff, R. F., Tudor, G. E., Pieper, K. S., & Hasselblad, V. (2006). Can One Assess Whether Missing Data Are Missing at Random in Medical Studies? *Statistical Methods in Medical Research, 15*(3), 213-234.

36. Rahman, A. A., & Bennett, D. (2009). Advanced Manufacturing Technology Adoption in Developing Countries.*Journal of Manufacturing Technology Management, 20*(8), 1099-1118.

37. Rahman, A. A., Brookes, N. J., & Bennett, D. J. (2009). The Precursors and Impacts of BSR on AMT Acquisition and Implementation. *IEEE Transactions on Engineering Management, 56*(2), 285-297.

38. Raymond, L. (2005). Operations Management and Advanced Manufacturing Technologies in SMEs: A Contingency Approach. *Journal of Manufacturing Technology Management, 16*(7/8), 936-955.

39. Raymond, L., Bergeron, F., & Croteau, A.-M. (2013). Innovation Capability and Performance Of Manufacturing SMEs: The Paradoxical Effect of IT Integration. *Journal of Organizational Computing and Electronic Commerce, 23*(3), 249-272.

40. Raymond, L., & Croteau, A.-M. (2006). Enabling the Strategic Development of SMEs through Advanced Manufacturing Systems. *Industrial Management & Data Systems, 106*(7), 1012-1032.

41. Raymond, L., & Croteau, A.-M. (2009). Manufacturing Strategy and Business Strategy in Medium-Sized Enterprises: Performance Effects of

Strategic Alignment. *IEEE Transactions on Engineering Management, 56*(2), 192-202.

42. Raymond, L., & St-Pierre, J. (2005). Antecedents and Performance Outcomes of Advanced Manufacturing Systems Sophistication in SMEs. *International Journal of Operations & Production Management, 25*(5/6), 514-533.

43. Schafer, J. L., & Graham, J. W. (2002). Missing Data: Our View of the State of the Art. *Psychological Methods, 7*(2), 147-177.

44. Small, M. H. (2007). Planning, Justifying and Installing Advanced Manufacturing Technology: A Managerial Framework. *Journal of Manufacturing Technology Management, 18*(5), 513-537.

45. Sohal, A. S., Sarros, J., Schroder, R., & O'Neill, P. (2006). Adoption Framework for Advanced Manufacturing Technologies. *International Journal of Production Research, 44*(24), 5225-5246.

46. Spanos, Y. E., & Voudouris, I. (2009). Antecedents and Trajectories of AMT Adoption: The Case of Greek Manufacturing SMEs. *Research Policy, 38*(1), 144-155.

47. Tambe, P., & Hitt, L. M. (2012). The Productivity of Information Technology Investments: New Evidence from IT Labor Data. *Information Systems Research, 23*(3), 599-617.

48. Thomas, A. J., Barton, R., & Chuke-Okafor, C. (2009). Applying Lean Six Sigma in a Small Engineering Company – A Model for Change. *Journal of Manufacturing Technology Management, 20*(1), 113-129.

49. Thomas, A. J., Barton, R., & John, E. G. (2008). Advanced Manufacturing Technology Implementation.*International Journal of Productivity and Performance Management, 57*(2), 156-176.

50. Thomas, A. J., & Barton, R. A. (2012). Characterizing SME Migration towards Advanced Manufacturing Technologies. *Proceedings of the Institution of Mechanical Engineers, Part B: Journal of Engineering Manufacture, 226*(4), 745-756.

51. Uwizeyemungu, S., & Raymond, L. (2012). Impact of an ERP System's Capabilities upon the Realisation of its Business Value: A Resource-Based Perspective. *Information Technology and Management, 13*(2), 69-90.

52. Withers, M. C., Drnevich, P. L., & Marino, L. (2011). Doing More with Less: The Disordinal Implications of Firm Age for Leveraging Capabilities for Innovation Activity. *Journal of Small Business Management, 49*(4), 515-536.

53. Zhang, Q., Vonderembse, M. A., & Cao, M. (2006). Achieving Flexible Manufacturing Competence. *International Journal of Operations & Production Management, 26*(6), 580-599.

54. Zhou, H., Leong, G. K., Jonsson, P., & Sum, C.-C. (2009). A comparative Study of Advanced Manufacturing Technology and Manufacturing Infrastructure Investments in Singapore and Sweden. *International Journal of Production Economics, 120*(1), 42-53.

Chapter 2

EFFECTIVE FACTORS ON ADVANCED MANUFACTURING TECHNOLOGY IMPLEMENTATION PERFORMANCE: A REVIEW

S. Saberi[1], R. Mohd. Yusuff[1], N. Zulkifli[1], and M.M.H. Megat Ahmad[2]

[1]Department of Mechanical and Manufacturing Engineering, Faculty of Engineering, University Putra Malaysia, 43400 UPM Serdang, Selangor, Malaysia

[2]Faculty of Engineering, University Pertahanan Nasional Malaysia, Kern Sungai Besi. 57000 Kuala Lumpur, Malaysia

ABSTRACT

This study reviews an extensive body of literature to investigate the factors effective on performance of companies implementing Advanced Manufacturing Technology (AMT). The purpose of this study is to provide a comprehensive viewpoint of issues related to successful AMT implementation and offer some directions to managers and investigators to make a company well-prepared to accept technology. The factors are grouped into three categories: technological, organizational and internal/external. The literature showed that in order to have a fruitful result from AMT investment, the organizational structure and culture, operational strategy and human resource should be organized and integrated appropriately with each other to avoid probable barriers and problems. Proposed framework can be used as a guideline for managers and investors in improving their AMT implementation process.

INTRODUCTION

A variety of pressures either locally or globally encourage manufacturers to become more agile, responsive and flexible if they wished to survive (1994). Firms that operate in developing, and/or newly industrialized countries face many uncertainties when venturing into the modern global markets (Noori, 1997). Thus, it was vital for manufactures to have the ability to compete due to the globalization in all aspects of product manufacturing such as product variations, labor, technology and markets (Mitala and Pennathur, 2004). These included massively increased competition and globalization of manufacturing and they served to place emphasis on a wide set of nonprime factors such as design, product innovation frequency, customization and delivery responsiveness (Bessant, 1994).

These conditions are bringing great challenges to firms, which can affect corporate strategic directions and alter business and manufacturing strategies. In an effort to survive under such conditions, companies are giving a strategic role to manufacturing, from simply supporting marketing strategies to playing a major role in strengthening a company's market position (Monge et al., 2006). The effective implementation of advanced manufacturing technology is considered to overcome this turbulent and hostile environment. This option is an important solution especially for small and medium size companies (Rosnah et al., 2003) in which lack organic structure and inadequate level of skilled workers and engineers and are not aware of the ways in which AMT can be helpful for them (Yusuff et al., 2005).

The rapid growth in both availability and range of AMT choice opens up major opportunities not only for improving substitution innovation but also for radical alternatives. These opportunities have never been done before and are doing in ways which were not possible hitherto (Bessant, 1994). Changes in communication and interaction related to AMT implementation have been shown to result in greater satisfaction with the technology (Stock and McDermott, 2001) and AMT adoptions appear to be a key condition for long term competitiveness. However, many AMT projects fail to meet the expectations of their adopters (Koc and Bozdag, 2009) and increasing signs of difficulty began to emerge which suggested that the translation of potential benefits into real competitive advantage was not always as simple as signing a cheque for a new piece of equipment (Bessant, 1994). In many cases not only AMT investments have been criticized for not yielding the desired results (Chung, 1996), but also some researchers found a negative contribution of AMTs to the firm performance (Boyer et al., 1997; Swamidass and Kotha, 1998). The researchers concluded that the relationship between AMTs and firm performance has a complex relationship (Koc and Bozdag, 2009) and the link is influenced by other factors, some controllable and some not controllable (Heine et al., 2003).

Thus, applying and adopting new technologies indicated that there are broader issues that have to be considered. Management of firms that are considering the adoption of AMT need to recognize, understand and address these issues in order to overcome or circumvent the problems of previous installations. They require knowing what the organizational and strategic factors are which make a firm more competitive and adept at using AMT in improving its performance and whether AMT's impact on company performance more pronounced if associated with a compatible organizational design and human force and management practices. Because of high cost and moderate-to high risk involved in AMT investment, it is so important for any organization to know more about these the factors. Generally, the investigated factors can be classified as technological, organizational and internal/external. This classification is illustrated in Fig. 1.

This study is a step in paving the way to provide an overview and guidance in AMT adoption and the right mix of strategic and important elements that leads to effective use of AMT in enhancing company performance.

Overview of AMTs

In studying AMT implications, the choice of AMT types and their classification is a decision of crucial importance that should be made on the basis of existing theory and the nature of the research study to be conducted. Advanced manufacturing technology has different meanings in different situations, but it can be broadly defined as 'an automated production system of people, machines and tools for the planning and control of the production process, including the procurement of raw materials, parts, components and the shipment and service of finished products (McDermott and Stock, 1999). More specifically, AMT can be described as a group of computer-based technologies, including Computer-Aided Design (CAD), robotics, Flexible Manufacturing Systems (FMS), Automated Materials Handling Systems (AMHS), Computer Numerically Controlled (CNC) machine or other automated identification techniques (Small and Yasin, 1997a).

Youssef (1992) defined advanced manufacturing technology as a group of integrated hardware and software based technologies, which if properly implemented, controlled and evaluated, will improve the efficiency and effectiveness of the firm. Boyer et al. (1997) used the term advanced manufacturing technology in their research to describe a variety of technologies like CAD and Electronic Data Interchange (EDI) which primarily utilize computers to control, track, or monitor manufacturing activities, either directly or indirectly. In addition, several technologies or programs such as bar codes or group technology which do not directly involve computers are also considered to be AMTs since they are closely associated with other AMT technologies.

AMT has been classified in different ways. Based on the automation and integration of manufacturing activities, Ghani and Jayabalan (2000) and Ghani et al. (2002) have been classified AMT into four levels.

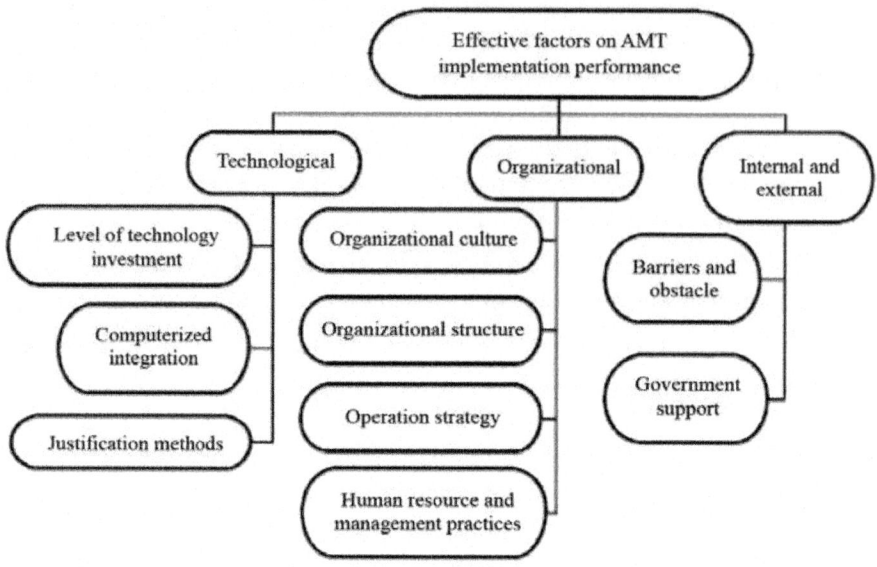

Figure 1: Contextual factors effective on company performance

First level includes numerically controlled machine and robots called stand-alone machine tools or equipments that are controlled by self-contained computers. In level 2 or manufacturing cells a grouping of machines such as group technology and flexible manufacturing system perform a variety of tasks to produce a family of parts. In level 3 cells in level 2 are connected to form linked islands through network of computerized information like computer-aided design/computer-aided manufacturing, automated storage and retrieval systems. In level 4 all the manufacturing activities including marketing of products are integrated through information network and formed computer-integrated manufacturing.

Waldeck (2007) classified advanced technologies in her study by Factor analysis in two levels: first level is Basic technology including Computer-aided design, Computer-aided manufacturing (CAM) and Direct numerical control. Next level is Artificial intelligence or complex technologies comprising vision systems, knowledge-based systems and decision-support systems. Zhang et al. (2006) also classified AMTs as Design technologies, such as CAD and CAE that support product design and engineering; Manufacturing technologies, such as CNC, CAM and AMHS which make production easier and faster; Planning and control activities are facilitated by the development of MRP, MRP II, electronic data interchange and bar coding and Integration technologies such as CIM, Local Area Networking (LAN) and enterprise-wide resource planning that allow a flow of information and coordinated decision-making between functions within and between the firms.

Small and Chen (1997) and Small and Yasin (1997b) classified AMT into three levels based on complexity, automation and integration of manufacturing activities.

Stand-alone systems include machine tools or equipment controlled by independent computers such as (computer-aided design and computer-aided process planning (CAPP)); Intermediate systems contain a group of machines to produce a family of parts such as (automated guided vehicles (AGVS) and automated storage and retrieval systems (AS/RS)); and Integrated systems which are connected to form linked islands through computerized information network, for example (flexible manufacturing systems and MRP). In summary, Table 1 presents the classifications found in the literature.

It is reasonable to state that most technological advancements that have changed the nature of manufacturing performance have taken place since 1950 (Mitala and Pennathur, 2004). AMTs enable both economies of scale and economies of scope to be achieved without changing the hardware and allow firms to blend small-batch and custom-order operations with the low-cost efficiency of standardized mass production (Efstathiades et al., 2002). The major strategic benefits that these technologies offer are the increased flexibility and responsiveness, enabling an organization to improve substantially its competitiveness in the marketplace (Millen and Sohal, 1998; Efstathiades et al., 1999). AMT has been viewed as a strategic weapon to gain competitive advantage, to improve productivity and performance, to enhance quality of production (Zhao and Co, 1997; Efstathiades et al., 2002) and decrease lead-time (Preece, 1995; Ghani and Jayabalan, 2000; Hofmann and Orr, 2005). In effect AMT changes the external risk propensity of the firm from risk-averse to risk-prone. That is, firms using AMT in practice create a series of call options to enter new markets and industries in the future (Efstathiades et al., 1999). It also was mentioned that even the benefits of advanced techniques such as Just-In-Time can be realized with applying only a few component of JIT and as a result companies can gradually invest in these technologies to get the most benefit from it (Yusuff et al., 1997).

Certainly, it takes some time for plants to realize the potential benefits of an AMT investment. It can be because of the learning curve associated with these technologies that may delay performance gains. As a fairly complicated technology, employees need extensive training and experience to master for new technologies. Therefore, time may act as a confounding variable in obtaining AMT benefits (Boyer, 1999; Nahm et al., 2006).

Table 1: Advanced manufacturing technology classification

Resource(s)	Dimensions
Boyer *et al.* (1996), Jonsson (2000), Dýaz *et al.* (2003)	Design, Manufacturing, and administrative
(Swamidass and Kotha, 1998)	Information exchange and planning technology, Production design technology, High-volume automation technology, and low-volume flexible automation technology
Small and Chen (1997) Small and Yasin (1997a, b)	Stand-alone, Intermediate, and integrated systems
Sanchez (1996), Beaumont *et al.* (2002)	Direct, indirect, and administrative
Meredith (1987)	Engineering techniques, manufacturing techniques, business techniques
Ghani and Jayabalan (2000), Ghani *et al.* (2002)	Stand-alone, manufacturing cells, linked islands, integrated manufacturing
Majchrzak and Paris (1995)	Integrated AMT , Non-integrated AMT
Kotha (1991), Kotha and Swamidass (2000)	Product design technologies, process technologies, logistics/planning technologies, information exchange technologies
Beaumont and Schroder (1997)	Direct, indirect, communication
Zhang *et al.* (2006)	Design technologies, manufacturing technologies, planning and control, integration technologies
Waldeck (2007)	Basic technology, artificial intelligence
Small (2006)	Stand-alone, moderate, and high complexity
Burgess and Gules (1998)	Hard technologies, soft technologies

A Synopsis of Performance Measurement

Evaluating the performance of AMTs relies on defining what success means (Burgess et al., 1997). For a company, performance is a measure of where it is; how far it has achieved its per-specified plans and more importantly, how it can efficiently use its capacity to improve its performance compared with its competitors (Agarwal, 1997).

At the beginning of the 1980s, AMT was seen only as a panacea to solve the financial problems in manufacturing companies. Managers concentrated only on the financial measures (Kidd, 1990) such as sales growth, market share and return on investment to justify the AMTs. Later on, the researchers found that such criteria do not capture the information that is required to judge the true effectiveness and outcomes of the new technologies. They found that the use of AMTs has substantial impacts to both individual and process requirements as the processes are reconfigured through computerization. They focused on operational measurement like the productivity and flexibility to justify the purchase of equipment to upper management. Organizational measures include other company's performance criteria like workflows, communication, integration of work and managerial control, also were considered as a new measurement system for AMT outcomes (McDermott and Stock, 1999; Efstathiades et al., 2002). Those measurement criteria were not enough to assure a manger whether his company got all possible benefits from AMT or not. AMTs are in relation with increased job responsibilities and the creation of new roles for employees. These technologies might enlarge employee control, even though such control may be limited to task design as opposed to task execution (Siegel et al., 1997). Therefore, human resource benefits relate to human force such as operator autonomy and the use of work teams were introduced (Chung, 1996; Mitala and Pennathur, 2004; Waldeck and Leffakis, 2007).

EFFECTIVE FACTORS ON AMT IMPLEMENTATION PERFORMANCE TECHNOLOGICAL CHANGE

Integration

One of the greatest advantages associated with AMTs is that of integration. Theoretically, by using the abilities of computers to electronically connect different machines and workstations together, a single integrated system will be formed to control all of the activities of a given firm starting with raw materials and finishing with finished goods ready to deliver to the final customer (Boyer, 1994). Integration either realized through computer-integrated transactions between functions, for example between marketing, engineering, production and maintenance, or between processes, such as linked between product design (e.g., CAD) with Process Planning (e.g., CAPP), manufacturing (like CNC), or production planning (e.g., MRP II) (Jonsson, 2000). Much of the attraction of AMTs has always been the potential to integrate different systems to create a complete system in which information and production can be controlled by computer, without the need for considerable human intervention (Boyer, 1994).

From an organizational perspective, Nemetz and Fry (1988) and Parthasarthy and Sethi (1992) argued that integration have resulted from using AMTs. Boyer et al. (1996), Diaz et al. (2003) and Melnyk and Narasimhan (1992) believed that more investment of AMTs leads to heavier integration between processes. Jonsson (2000) showed that the level of investment in AMT and the integration between processes/functions affect the performance of companies. It is concluded that technology integration offers more benefits than the automation of individual processes (Boyer, 1994).

Justification Approaches

Obviously the first decision any firm considering in AMTs adoption is whether such an investment would be wise (Boyer, 1994). The high costs of hardware and software of many of the advanced manufacturing technologies and the complexities of the operational and organizational problems related to adopting these systems, make justification a necessary but difficult proposition (Small, 1993). Generally justification methods has been classified into three categories of approaches: economic, analytic and strategic approaches (Suresh and Meredith, 1985). Economic methods justify based on the cost reduction or capacity expansion (Mcdaniel, 1989; Small, 1993). The Analytic methods frequently consider uncertainty, flexibility, risk and non-economic benefits of AMTs (Meredith and Suresh, 1986; Mcdaniel, 1989; Chan et al., 2001). The strategic approaches tend to be less quantitative than either the economic or analytic techniques and typically involve subjective estimates of the key indicators or surrogate measures related to strategic objectives (Mohanty and Deshmukh, 1998; Chan et al., 2001). It is demonstrated that most companies are using only one method, most probably simple economic method or using hybrid evaluation approaches (i.e., strategic and economic or economic and analytic) (Meredith and Suresh, 1986; Chadwell-Hatfield et al.,1996; Stuart, 2002) for AMT investment decision but Small and Chen (1997) showed that companies utilizing hybrid method attained higher levels of success from their AMT projects than plants that used only one method. It seems that inappropriateness of one criterion might be partly balanced by the use of the other methods. It is unlikely that any single justification method will lead companies to all or even a wide range of AMT benefits and improve performance. Thus, integrated approaches (i.e. using strategic, economic and analytic methods in parallel) were recommended to quantify the tangible and intangible benefits throughout the technology investment (Small and Chen, 1995).

Organizational Design

Successful implementation of AMT involves the mutual adaptation of both the new technology to the organization and the organization to the technology (Frohlich, 1998). In fact the adjustment of technology to the organization and vice versa can ease the accomplishment of new technologies and avoid management problems associated with AMTs (Yusuff et al., 2004a). This importance embraces structure, culture and strategy of any organization.

Organizational Structure

It has been argued that manufacturing companies that adopt AMTs without first redesigning organizational structures and processes, encountering high difficulties (Millen and Sohal, 1998). Along with AMTs emergence, industrial organizations have deeply changed their manufacturing processes through the acquisition of computerized technologies. This evolvement is frequently viewed as the basis for a new industrial revolution-the arrival of the factory of the future- and new form of organizational structure (Dean et al., 1992). Generally, structure of an organization is the formal system of working relationships that share and harmonize the tasks of multiple people and groups to serve a common purpose. Centralization, formalization and complexity are the three dimensions often use in research and practice to describe structure. Centralization in the organization refers to the delegation of power among the jobs. The less power delegated in an organization the greater the centralization in the organization and vice versa. Formalization refers to the extent to which expectations regarding the aims and objectives of work are specified and written. Highly formalized organization structures recommend what each individual should act based on rules and procedures that are obtainable. Last dimension, Complexity, refers to the number of distinctly different job titles or occupational groupings and the number of definitely dissimilar units/ departments, in a group/organization (Gibson et al., 1973).

The structure of the organization has been considered as the key factor to successfully implementing AMT in various literatures (Dalton et al., 1980; Kotha, 1991; Dean et al., 1992; Belassi and Fadlalla, 1998; Ghani et al., 2002; Jin-Bo et al., 2006; Song et al., 2007; Sun et al., 2007). It is theorized that the correct organizational structure is in place, a company will be more successful in implementing advanced manufacturing technologies (Boyer, 1994; Anderson, 1998). Boyer et al. (1996) stated that the multiple levels of authority involved with hierarchical organizations often represent an obstacle to the effective implementation of AMTs and streamlining the organization with fewer level of authority brings a greater ability to integrate AMTs. They concluded that rigid, bureaucratic organizational structure which has been associated with highly automated, but non-computerized manufacturing systems such as assembly lines, is not appropriate for more flexible technologies. Gupta et al. (1997) also indicated that only decentralization with fewer rules and more employee involvement were positively relate to technology whereas formalization and mechanistic structure interacted negatively with AMT. The result of this study emphasized that irrespective of the technology type, a firm needs to be as least mechanistic as possible to be effective. In examining the relationship between structure and AMT Ghani et al. (2002) found that, at high proactive level, the mechanistic structure of AMT plants has been found to change into an organic structure. In fact organizations with many different types of jobs and departments generate more complicated managerial and organizational problems than those with fewer jobs and units. Flatter, less complex structures with maximum administrative decentralization, are more likely toward creating a potential for improved attitudes, more effective supervision, greater individual responsibility and company performance (Belassi and Fadlalla, 1998; Malhotra et al., 2001; Chang and Lung, 2002).

Organizational Culture

Successful implementation of AMT often requires dissimilar types of organization and or management practices than are found in more traditional environments (Zammuto and O'Connor, 1992). This is because new technologies directly challenge established norms and strategic options. Organizational culture referred to a holistic construct that describes the complex set of knowledge structures which organization members use to perform tasks and generate social behavior. This construct is affected by and impacts many aspects of organization such as structure, role expectations and job description. Culture defines how to act on the job, who makes decision in various situations and how to think and behave toward coworkers, supervisors, industry norms and practices. This view of culture includes the organization's internal system of power including formal authority structures, control systems, task structures and organization rules (Bates et al., 1995). In other word, culture is to the organization what personality is to the individual, a hidden, yet unifying topic that provides meaning, direction and mobilization (Belassi and Fadlalla, 1998).

Generally the culture was picture into two main dimensions as flexibility and control (Zammuto and O'Connor, 1992; Denison and Mishra, 1995). Flexibility-oriented culture is based on norms and values related to the affiliation. It focuses on the development of human resources and values member involvement in decision making. In this culture, individuals are encouraged by the significant or ideological appeal of the task being undertaken. On the other hand, control-oriented culture is penetrated by assumptions of achievement such as planning, productivity and efficiency. More specifically, assumptions of stability are the foundation of this culture and individuals respect to the organizational mandates because roles are formally announced and enforced through rules and regulations.

Regarding to the effect of culture on company performance, Zammuto and O'Connor (1992) hypothesized that the control-oriented approach may well lead to increased productivity, but can hinder AMTs implementation, because centralization of responsibilities diminishes opportunities for organizational learning, which, in turn, can make more difficulties to get an AMT up and running reliably. They concluded that flexibility-oriented values will gain AMTs' productivity and flexibility benefits. McDermott and Stock (1999) examined how organizational culture is related to outcomes associated with advanced manufacturing technology implementation, such as, operational benefits, organizational or managerial benefits, competitive benefits and satisfaction. They found that implementation effects that may take longer to happen, such as overall satisfaction or competitive performance, did depend on the cultural flexibility. Chang (2000) tried to find the relationship between organizational culture and successful implementation of AMTs in Taiwan. Analysis showed that the control-oriented culture did result in reduced AMT implementation success while good internal process, rational goals and horizontal coordination have positive effect. Results demonstrated that companies with a history of successful AMT implementation preferred a flexibility-oriented culture in a more conducive environment to ease the AMT implementation (Yusuff et al., 2008).

Operational/Manufacturing Strategy

Basically, the importance of manufacturing strategy to the overall success of the corporation has received considerable attention as Skinner published his landmark article in 1969, manufacturing-missing link in corporate strategy (Mcdaniel, 1989). According to the new approach to manufacturing strategy, managers should think about investments more in their capacity to build new capabilities that provide enduring sources of competitive advantage and are usually built over time through a series of investments in facilities, human capital and knowledge. The early approach to manufacturing strategy led top managers to focus their companies' operations around specific competitive priorities that tended to make them vulnerable to strategic shifts. A good manufacturing strategy was one that defended a company's position through a narrowly focused set of capabilities (Hayes and Pisano, 1994). In other words, strategy was denoted as actions or patterns of actions intended for the achievement of goals. The term strategy covers more than just intended or planned strategy in an organizational setting; it also contains the sequence of decision that exhibit a posteriori consistencies in decisional behavior (Swamidass and Newell, 1987).

There is general agreement that a firm's operations/manufacturing strategy is comprised of four key competitive priorities: cost, quality, flexibility and dependability/delivery (Nemetz and Fry, 1988; Dangayach and Deshmukh, 2003). Cost strategy is based on the production and distribution of product at lower cost. It is a measure of the manufacturing function's efficiency and traditionally it has been associated with high volume/mass production. Quality strategy is associated with a firm's ability to provide superior products or services, often at higher prices. Dependability/Delivery strategy is defined with on-time delivery schedules and quickly response to customer orders. Flexibility is a measure of a firm's ability to react to market demands by switching from one product to another through matched policies and actions and react to changes in production and product mix, modifications in design, fluctuations in materials and changes in sequence. The effectiveness of a company's operations strategy is the function of degree of linkage or consistency between the competitive priorities that are emphasized on the corresponding decisions regarding the structure and infrastructure of operations (Hayes and Wheelwright, 1984; Boyer and Pagell, 2000; Stock and McDermott, 2001). The emphasis placed on these priorities varies by firms, depending on a large number of factors including availability of resources, business strategy, existing capability, managerial behavior, nature and intensity of competition and environmental condition (Agarwal, 1997).

One of company's most important variables for coping with environmental uncertainty is flexibility that is particularly relevant to the rapidly changing conditions affecting manufacturing organizations (Nemetz and Fry, 1988). This is nowhere more truly than for AMT, which provides the biggest source of flexibility in any manufacturing organization (Boyer and Pagell, 2000; Stock and McDermott, 2001). It has been noted that although AMT creates a world of opportunities, they will not be converted to advantage unless the adopting firm uses a strategic planning approach (Small and Yasin, 1997b). Swamidass and Newell (1987) conducted an empirical study to find the relationship between operational strategy and performance. They found that environmental uncertainty such as manufacturing flexibility and the role of manufacturing managers in strategic decision making influenced manufacturing strategy and among different dimension of manufacturing strategy, flexibility has a strong relationship with business performance. Efstathiades et al. (1999) declared that AMT implementation is more related to quality and delivery strategy. Results from Lewis and Boyer (2002) showed that among the two groups in their sample (high performers and low performers) in applying AMTs during the last 3 years, high performers generally were more likely to stress on flexibility, quality and delivery strategy than cost strategy and specifically the most dramatic difference appeared with regard to quality strategy between the two groups. Other researchers believed that all four manufacturing strategy dimensions are important in implementing new technologies and gaining related benefits (Ferdows and De Meyer, 1990; David et al., 1996) and focusing on one dimension does not relate directly to AMT performance. The simultaneous achievement of cost, quality, delivery and flexibility by many Japanese companies has highlighted this new possibility that can be realized by adopting advanced process technologies and management techniques (Agarwal, 1997).

Human Resource and Management Practices

Along with technology development, the human resource is an asset for any organization, without which the use and development of technology will not happen (Efstathiades et al., 2000) and has significant impact on strategic success (Malhotra et al., 2001). Human resources' qualities, attitudes and behavior can provide the firm with a source of competitive advantage with respect to its rivals (Bidanda and Cleland, 1995; Bayo-Moriones and De Cerio, 2004). Researchers emphasize the importance of providing appropriate workforce development activities such as socialization ability (Chen et al., 2008) and managers involvement in R and D projects (Liu and Tsai, 2007) to improve skills and relational requirements resulting from modifications in technology and new production processes in enhancing company performance. Inherently, all AMTs will increase worker requirements as workers are given more autonomy over issues including planning and problem solving (Waldeck, 2007). Evidence from the literature suggested that planning and implementation activities aimed at preparing workers for AMT adoption, play a critical role in guaranteeing an exploitation of the system benefits (Small and Yasin, 1997a, b). Thus, a major challenge for future successful implementations lies in addressing the needs imposed by AMT on the human elements.

In order to turn workers into key elements for building a competitive edge, people have to be managed in a distinctive way (Bayo-Moriones and De Cerio, 2004) and being more capable in terms of knowledge, skills, attitudes and responsibility (Bidanda and Cleland, 1995; Waldeck, 2007). As a result, providing workers with opportunities to improve their inherent motivation and job satisfaction by means of employee-involvement practices could be deemed an acceptable policy to ally the goals of employees with the firms using AMTs (Bessant, 1994; Bayo-Moriones and De Cerio, 2004; Waldeck, 2007). Education and training are also crucial to the successful implementation of AMT. Experience has shown that between 25 to 40% of the total cost of an extensive successful automation project should be spent on education and training (Zhao and Co, 1997). Firms with successful AMT implementations also enlist champions. These individuals support a continual driving force throughout the initiative (Millen and Sohal, 1998). The effect of these three practices beside the other seven factors were tested on ERP systems in some Malaysian companies and the results showed their importance in real situations (Jafari et al., 2009). Widening of the marketplace, increasing importance of technology and imperative of innovation and focusing on cross-functional groups, are appropriate ways to develop viable business solutions (Doolen et al., 2003). Because of the importance of the management personal characteristics, experience and background on their decisions, any change has to start with the managers on the top and in the middle, then the organization of workers on the shop floor (Sun and Gertsen, 1995). Besides, to facilitate the psychological, physical and cultural change resulting from AMT implementation, management must build trust and co-operation (Cook and Cook, 1994). These practices can grantee the achievement of technology investment projects.

Internal/External factors

Government Support

An important feature of the optimistic climate common in the 1980's was a strong faith placed by government relate to AMT affecting manufacturing industry. It is generally recognized that some measure of state support for innovation is necessary in order to preserve a position of international competitiveness. During this period, governments of most advanced industrial economies proposed a range of programs designed to smooth the progress of advanced manufacturing technologies adoption (Bessant and Rush, 1993). Approximately, the increase in interventionist-technology policy-making has been the outcome of increasing government awareness of the extent that innovation is joined to economic growth and with the recognition of the strategic role of technologies in the evolvement of new industries and markets (Vickery and Blau, 1989; Hilpert, 1991). Government support programs was introduced to compensate for deficiencies in economic/industrial environment, to strengthen the technological infrastructure that facilitates the convey of technology from the developer or supplier to the user, to tackle specific firm-level obstacles of technology diffusion and to increase the supply of technical and managerial personnel (Bessant and Rush, 1993). It is argued that end-users of AMT, mostly private firms in the manufacturing sector, are competing in the global market with leading foreign companies and hence in need of up-to-date AMT. Beside, domestic and AMT suppliers and R and D institutes lack the native capability to meet the complex demands. It is not unusual that technology developers and technology users are separated from each other. Meanwhile, governments programs coordinate these diverse and even contradictory demands in developing national policy of AMT (Park, 2000). Lay (1993) analyzed how subsidized firms have been planning and implementing their technology (CIM, in that case) and whether AMT projects in subsidized firms differ, from those in non-subsidized firms. For about two-thirds of supported firms, public support for a technology project showed the effect of extending or speeding up the finalizing of the project already planned to take place at that time. A group of firms stated that they would not have embarked on the project at all, or not at that time, without public support. In other words, these programs including direct financial support, information/consultancy support and so on, affect the competitiveness of firms in promoting their market share, open up significant new opportunities and allow firms to deal with strategic challenges in their environment (Lay, 1993).

Barriers/Obstacles

By the mid-1980s the diffusion of AMT, as measured by adoption rates, was high amongst large firms and trickling down to the small and medium sized companies. However increasing signs of difficulty began to emerge which suggested that the translation of potential benefits into real competitive advantage was not always as simple as signing the check for a new piece of equipment (Bessant, 1994). Successful implementation of AMT projects needed persistent efforts to integrate operating and organizational systems to support these operations (Small, 1993). The introduction of a new technology can reduce performance when the organization initially struggles to acquire the requisite skills and knowledge; i.e., there may be an substantial lag between installing new technology and getting benefits from it (Beaumont and Schroder, 1997).

Beatty and Gordon (1990) list three classifications of barriers to implementation: structural, related to organizational infrastructure and justification difficulties; human, related to uncertainty and workers' resistance; and technical, related to the incompatibility of systems. Adler (1988) suggests that in the decisive majority of cases, the human resource management issues are the major stumbling blocks in implementing the new technologies. Meredith (1987) notes that in the early stages of FMS implementation, human and organizational infrastructure and worker education were major difficulties. Bessant (1994) concluded that AMT investment was unlikely to succeed unless it was located within a coherent business strategy and accompanied by relevant parallel organizational change. Sambasivarao and Deshmukh (1995) showed that adoption of AMTs involves major investment and a high degree of uncertainty and hence, warrants significant consideration within a manufacturing firm at the strategic level. In a survey among Malaysian small and medium size companies, Yusuff et al. (2004b) identified that lack of understanding of technologies and inappropriate planning are the biggest obstacles in obtaining the strategic benefits of AMT implementation. Shortage of suitable man power (Cook and Cook, 1994; Zhao and Co, 1997; Shepherd et al., 2000), inadequate organizational planning and preparation for the adoption of the AMT (Small and Yasin, 1997a, b), failure to balance investments in technological systems with investments in the infrastructure to support these systems (Boyer et al., 1997), inadequate cost-justification methods (Cook and Cook, 1994), technology mania, lack of top management's continued support, financial limitation (Ratnasingam et al., 2009) and inadequate managerial training for AMT projects (Marri et al., 2007) are other major problems hindering the success of factory automation. In summary Table 2 illustrates all explored factors and their dimension and related references.

Table 2: A summary on effective factors

Factor	Factor dimension	Supported Literature
Advanced Manufacturing Technology	Stand-alone systems Intermediate systems Integrated systems	Small and Chen (1997), Small and Yasin (1997a, b), Small (2007)
Justification method	Economic justification Strategic justification Analytic justification	Suresh and Meredith (1985), Chan et al. (2001), Small (2006), Narain et al. (2007)
Integration		Boyer et al. (1996), Jonsson (2000), Diaz et al. (2003)
Manufacturing strategy	Flexibility strategy Cost strategy Delivery strategy Quality strategy	Lewis and Boyer (2002), Diaz et al. (2003)
Organizational structure	Centralization, Formalization Complexity	Gupta et al. (1997), Belassi and Fadlalla (1998)
Organizational culture	Flexibility Control	Zammuto and O'Connor (1992), Chang (2000)
Human resource and Management practices		Chung (1996), Small and Yasin (1997, 2000) Bessant (1994), Cook and Cook (1994), Majchrzak and Paris (1995)
Obstacles and Barriers		Cook and Cook (1994), Schroder and Sohal (1999), Hofmann and Orr (2005)
Government programs		Bessant and Rush (1993), Lav (1993), SMIDEC (2007)

The Interaction Effect of Variables

Research has indicated that the application of AMT can be successful if only designing technology, organization and people are base on the principle of reinforcing each other and their integration (Sun et al., 2007; Waldeck, 2007). As employees need higher knowledge/skill and organizations adopt teamwork gradually, decentralization among organization occurs (Gupta et al., 1997). In new AMT environments, employees are not only single operators, but would be coordinators or decision-makers. The role of organizations can not be a single task-distributor and coordinator anymore, but it would promote employees' enthusiasm and independence through multifold manners to bring them into the most potential (Sun et al., 2007). Fewer complexities in a flatter organization are helpful to encourage employees to apply AMT and enhance their responsibility. Less formalization could stimulate employees, awake their sense of responsibility and improve working efficiency of employees and implementation effects of AMT (Song et al., 2007). Consequently, organic structure with less complexity will be the feature of decentralized management, minimal organization levels and more teamwork enterprises that must reduce organization levels to make fast transfer of information and communication (Sun et al., 2007). Less formal delegation of authority in flexibility-oriented culture allows top management to provide the general strategic direction. Lower level management is then free to work and innovate under the assumption that its efforts will lead the organization towards the desired, top management imposed (Gupta et al., 1997).

In addition, effective cooperation between process change and factors of organizational change is good to the achievement of performances of AMT (Song et al., 2007). Manufacturing strategy is best implemented when plant personnel understand the strategic aims and direction of the plant and can exercise appropriate judgments in less formalized organization. Once again it is impossible to speculate about casualty but this association may indicate that processes of strategic goal orientation and decentralized decision making reinforce each other over time (Bates et al., 1995). Well aligned and implemented manufacturing strategy was found to coexist with a flexible-oriented organizational culture. It is indicated that a well arranged strategy, which includes informal planning processes, communication strategy and contribution to all four dimension of competitive priorities, coexist with- a clan oriented culture characterized by the use of group and teams, low emphasis on hierarchy and high level of loyalty and shared plant-wide philosophy. +Consequently, contribution of these factors leads companies materializing their wish by applying new advanced technologies.

But the performance story does not end to this point. By applying performance appraisal to measure the performance of the employees and the organization, firms are capable of checking the progress towards the desired goals and aims (Badawy, 2007). The history of performance appraisal can be dated back to the 20th century and then to the second world war when the merit rating was used for the first time. An employer evaluating their employees is a very old concept. Performance appraisals are an indispensable part of performance measurement.

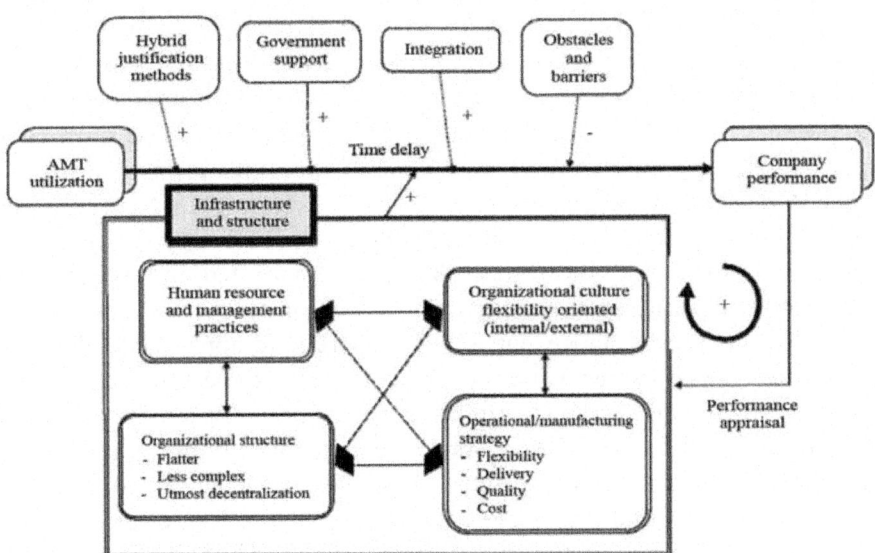

Figure 2: A framework of investigated factors and their relationship with company performance

It takes into account the past performance of the organization and focuses on the improvement of the future performance of the organization.

The performance appraisal process provides an opportunity for introducing organizational change. It facilitates the process of change in the organizational culture/structure. The interactive sessions between the management and the employees, the mutual goal setting and the efforts towards the career development of the employees help the organization to become a learning organization that can solve the obstacles better than the past (Gloet and Terziovski, 2004). Conducting performance appraisals on a regular basis helps it to become an ongoing part of everyday practice and helps employees to take the responsibility of their work and boosts their professional development and as a result, company productivity. Totally Fig. 2 shows a framework containing all investigated elements and their relationship with the company performance.

CONCLUSIONS

Today, technological capabilities can be strategically used to achieve sustainable competitive advantage and the implementation of these technology is an organizational transformation process, in which people's value, organizational culture, competition strategy and arrangement of people all will change to well-matched with each other (Zhao et al., 1992). The key to successful AMT implementation appears to be the collaboration of appropriate factors and their integration that will offer maximum benefits from AMT implementation.

The meaning of this research is to provide a comprehensive study on that systematically builds upon past researches in order to guide investigation into the successful AMT implementation and to determine those most critical organizational and strategic elements which if present can make a firm able to use AMT in enhancing performance. The framework presents the intra/inter-relationship among the variables influencing company performance in parallel with technology utilization that can be analyzed and offers testable propositions. This literature suggests that utilization of AMT will not also ipso facto guarantee performance but will further require appropriate changes in the firm's structure and infrastructure and continue with performance appraisal to improve company capability. In conclusion, the proposed framework can be used as a guideline for managers and engineers in improving their AMT implementation process. The offered framework is in the hope that it will stimulate empirical and practical investigations that will in turn generate the empirical evidence on which more adequate and strong hypothesis can be built.

REFERENCES

1.   Adler, P.S., 1988. Managing flexible automation. Calif. Manage., 30: 34-57.

2. Agarwal, D., 1997. An empirical investigation of the impact of advanced manufacturing technology on business performance. Ph.D. Thesis, City University of New York, pp: 248.

3. Anderson, B.K., 1998. An idiographic study of the relationship between organizational structure and the successful implementation of advanced manufacturing technologies in those companies exhibiting best practices. Ph.D. Thesis, The University of Alabama in Huntsville, pp: 170.

4. Badawy, M.K., 2007. Managing human resources. Res. Technol. Manage., 50: 56-74.

5. Bates, K.A., S.D. Amundson, R.G. Schroeder and W.T. Morris, 1995. The crucial interrelationship between manufacturing strategy and organizational culture. Manage. Sci., 41: 1565-1580.

6. Bayo-Moriones, A. and J.M.D. De Cerio, 2004. Employee involvement: Its interaction with advanced manufacturing technologies, quality management and inter-firm collaboration. Hum. Factors Ergon. Manuf., 44: 117-134.

7. Beatty, C. and J.R.M. Gordon, 1990. Advancd manufaturing technology: Making it happen. Bus. Q., 54: 46-53.

8. Beaumont, N., R. Schroder and A. Sohal, 2002. Do foreign-owned firms manage advanced manufacturing technology better. Int. J. Operat. Prod. Manage., 22: 759-771.

9. Beaumont, N.B. and R.M. Schroder, 1997. Technology, manufacturing performance and business performance amongst Australian manufacturers. Technovation, 17: 297-307.

10. Belassi, W. and A. Fadlalla, 1998. An integrative framework for FMS diffusion. Omega, 26: 699-713.

11. Bessant, J. and H. Rush, 1993. Government support of manufacturing innovations: Two country-level case studies. IEEE Trans. Eng. Manage., 40: 79-91.

12. Bessant, J., 1994. Towards total integrated manufacturing. Int. J. Prod. Econ., 34: 237-251.

13. Bidanda, B. and D.I. Cleland, 1995. Human issues in technology implementation-Part 1. Ind. Manage., 37: 22-26.

14. Boyer, K. and M. Pagell, 2000. Measurement issues in empirical research: Improving measures of operations strategy and advanced manufacturing technology. J. Operat. Manage., 18: 361-374.

15. Boyer, K.K., 1994. Patterns of advanced manufacturing technology implementation: Technology and infrastructure. Ph.D. Thesis, The Ohio State University, pp: 262.

16. Boyer, K.K., 1999. Evolutionary patterns of flexible automation and performance: A longitudinal study. Manage. Sci., 45: 824-842.

17. Boyer, K.K., G.K. Leong, P.T. Ward and L.J. Krajewski, 1997. Unlocking the potential of advanced manufacturing technologies. J. Operat. Manage., 15: 331-347.

18. Boyer, K.K., P.T. Ward and G.K. Leong, 1996. Approaches to the factory of the future An empirical taxonomy. J. Operat. Manage., 14: 297-313.

19. Burgess, T.F. and H.K. Gules, 1998. Buyer supplier relationships in firms adopting advanced manufacturing technology: An empirical analysis of the implementation of hard and soft technologies. J. Eng. Technol. Manage., 15: 127-152.

20. Burgess, T.F., H.K. Gules and M. Tekin, 1997. Supply chain collaboration and success in technology implementation. Integrated Manuf. Syst., 8: 323-332.

21. Chadwell-Hatfield, P., G. Bernard, H. Philips and A. Webster, 1996. Financial criteria, capital budgeting techniques and risk analysis of manufacturing firms. J. Applied Bus. Res., 13: 95-104.

22. Chan, F.T.S., M.H. Chan, H. Lau and R.W.L. Ip, 2001. Investment appraisal techniques for advanced manufacturing technology (AMT): A literature review. Integrated Manuf. Syst., 12: 35-47.

23. Chang, C.F., 2000. A study of the role of organizational values in advanced manufacturing technology implementation. DBA Thesis, Wayne Huizeng, Nova Southeastern University, pp: 72.

24. Chang, P.L. and S.S.C. Lung, 2002. Organizational changes for advanced manufacturing technology infusion: An empirical study. Int. J. Manage., 19: 206-217.

25. Chen, C.P., P.L. Liu and C.H. Tsai, 2008. A study of the influence of organizational knowledge ability and knowledge absorptive capacity on organization performance in Taiwan›s hi-tech enterprises. J. Applied Sci., 8: 1138-1148.

26. Chung, C.A., 1996. Human issues influencing the successful implementation of advanced manufacturing technology. J. Eng. Technol. Manage., 13: 283-299.

27. Cook, J.S. and L.L. Cook, 1994. Achieving competitive advantages of advanced manufacturing technology. Benchmark. Qual. Manage. Technol., 1: 42-63.

28. Diaz, M.S., J.A.D. Machucaa and M.I.J. Alvarez-Gil, 2003. A view of developing patterns of investment in AMT through empirical taxonomies: New evidence. J. Operat. Manage., 21: 577-606.

29. Dalton, D.R., W.D. Todor, M.J. Spendolini and L.W. Porter, 1980. Organization structure and performance: A critical review. Acad. Manage. Rev., 5: 49-64.

30. Dangayach, G.S. and S.G. Deshmukh, 2003. Evidence of manufacturing strategies in Indian industry. Int. J. Prod. Econ., 83: 279-298.

31. Dean, J.W., S.J. Yoon and G.I. Susman, 1992. Advanced manufacturing technology and organization structure: Empowerment or subordination? Org. Sci., 3: 203-229.

32. Denison, D.R. and A.K. Mishra, 1995. Toward a theory of organizational culture and effectiveness. Organiz. Sci., 6: 204-223.

33. Doolen, T.L., M.E. Hacker and E.M.V. Aken, 2003. The impact of organizational context on work team effectiveness: A study of production team. IEEE Trans. Eng. Manage., 50: 285-296.

34. Efstathiades, A., S. Tassou and A. Antoniou, 2002. Strategic planning, transfer and implementation of Advanced Manufacturing Technologies (AMT). Development of an integrated process plan. Technovation, 22: 201-212.

35. Efstathiades, A., S.A. Tassou, A. Antoniou and G. Oxinos, 1999. Strategic considerations in the introduction of advanced manufacturing technologies in the Cypriot industry. Technovation, 15: 105-115.

36. Efstathiades, A., S.A. Tassou, G. Oxinos and A. Antoniou, 2000. Advanced manufacturing technology transfer and implementation in developing countries: The case of the Cypriot manufacturing industry. Technovation, 20: 93-102.

37. Ferdows, K. and A. De Meyer, 1990. Lasting improvements in manufacturing performance: In search of a new theory. J. Operat. Manage., 9: 168-184.

38. Frohlich, M., 1998. How do you successfully adopt an advanced manufacturing technology? Eur. Manage. J., 16: 151-159.

39. Ghani, K.A. and V. Jayabalan, 2000. Advanced manufacturing technology and planned organizational change. J. High Technol. Manage. Res., 11: 1-18.

40. Ghani, K.A., V. Jayabalan and M. Sugumar, 2002. Impact of advanced manufacturing technology on organizational structure. J. High Technol. Manage. Res., 13: 157-175.

41. Gibson, J., J. Ivancevich and J. Donnelly Jr., 1973. Organizations: Behavior-Structure-Processes. Irwin, Homewood, IL.

42. Gloet, M. and M. Terziovski, 2004. Exploring the relationship between knowledge management practices and innovation performance. J. Manuf. Technol. Manage., 15: 402-409.

43. Gupta, A., I.J. Chen and D. Chiang, 1997. Determining organizational structure choices in advanced manufacturing technology management. Omega, 25: 511-521.

44. Hayes, R.H. and G.P. Pisano, 1994. Beyond world-class: The new manufacturing strategy. Harvard Bus. Rev., 72: 77-86.

45. Hayes, R.H. and S.C. Wheelwright, 1984. Restoring Our Competitive Edge: Competing Through Manufacturing. John Wiley and Sons, New York, USA.

46. Heine, M.L., V. Grover and M.K. Malhotra, 2003. The relationship between technology and performance: A meta-analysis of technology models. Omega, 31: 189-204.

47. Hilpert, U., 1991. State Policies and Techno-Industrial Innovation. Routledge, London.

48. Hofmann, C. and S. Orr, 2005. Advanced manufacturing technology adoption-the German experience. Technovation, 25: 711-724.

49. Jafari, S.M., M.R. Osman, M.Y. Rosnah and S.H. Tang, 2009. A consensus on critical success factors for enterprise resource planning systems implementation: The experience of Malaysian firms. Int. J. Manuf. Technol. Manage., 17: 396-407.

50. Jin-Bo, S., D. Da-Shuang and S. Yan-Qiu, 2006. The relationship between change of organizational structure and implementation of advanced manufacturing technology: An empirical study. Proceedings of the International Conference on Management Science and Engineering, Oct. 5-7, Lille, pp: 782-786.

51. Jonsson, P., 2000. An empirical taxonomy of advanced manufacturing technology. Int. J. Operat. Prod. Manage., 20: 1446-1474.

52. Kidd, P.T., 1990. Organisation, people and technology: Advanced manufacturing in the1990s. Comput. Aided Engmeer. J., 7: 149-153.

53. Koc, T. and E. Bozdag, 2009. The impact of AMT practices on firm performance in manufacturing SMEs. Robot. Comput. Integrated Manuf., 25: 303-313.

54. Kotha, S. and P.M. Swamidass, 2000. Strategy, advanced manufacturing technology and performance: Empirical evidence from U.S. manufacturing firms. J. Operat. Manage., 18: 257-277.

55. Kotha, S., 1991. Strategy, manufacturing structure and advanced manufacturing technologies. Academy of Management Best Paper Proceedings, pp: 293-297.

56. Lay, G., 1993. Government support of computer integrated manufacturing in Germany: First results of an impact analysis. Technovarion, 13: 283-297.

57. Lewis, M.W. and K.K. Boyer, 2002. Factors impacting AMT implementation: An integrative and controlled study. J. Eng. Technol. Manage., 19: 111-130.

58. Li, D., M.A. Hitt and J.D. Goldhar, 1996. Advanced manufacturing technology: Organizational design and strategic flexibility. Org. Stud., 17: 501-523.

59. Liu, P.L. and C.H. Tsai, 2007. The Influences of R and D management capacity and design/manufacturing integration mechanisms on new product development performance in Taiwans high-tech industries. J. Applied Sci., 7: 3628-3638.

60. Majchrzak, A. and M.L. Paris, 1995. High-performing organizations match technology and management strategies: Results of a survey. Int. J. Ind. Ergon., 16: 309-325.

61. Malhotra, M.K., M.L. Heine and V. Grover, 2001. An evaluation of the relationship between management practices and computer aided design technology. J. Operat. Manage., 19: 307-333.

62. Marri, H.B., Z. Irani and A. Gunasekaran, 2007. Advance manufacturing technology implementation in SMEs: A framework of justification criteria. Int. J. Elect. Bus., 5: 124-140.

63. McDermott, C.M. and G.N. Stock, 1999. Organizational culture and advanced manufacturing technology implementation. J. Operat. Manage., 17: 521-533.

64. Mcdaniel, J.R., 1989. Managing automation a study of the adoption, implementation and evaluation of advanced manufacturing technology. Ph.D. Thesis, University of Massachusetts.

65. Melnyk, S.A. and R. Narasimhan, 1992. Computer Integrated Manufacturing: Guidelines and Applications from Industrial Leaders. Irwin, Homewood.

66. Meredith, J.R. and N. Sureshm, 1986. Justification techniques for advanced manufacturing technologies. Int. J. Prod. Res., 24: 1043-1057.

67. Meredith, J.R., 1987. The strategic advantages of the factory of the future. Calif. Manage. Rev., 29: 27-41.

68. Millen, R. and A.S. Sohal, 1998. Planning processes for advanced manufacturing technology by large American manufacturers. Technovation, 18: 741-750.

69. Mitala, A. and A. Pennathur, 2004. Advanced technologies and humans in manufacturing workplaces: An interdependent relationship. Int. J. Ind. Ergon., 33: 295-313.

70. Mohanty, R.P. and S.G. Deshmukh, 1998. Advanced manufacturing technology selection: A strategic model for learning and evaluation. Int. J. Prod. Econ., 55: 295-307.

71. Monge, C.A.M., S.S. Rao, M.E. Gonzalez and A.S. Sohal, 2006. Performance measurement of AMT: A cross-regional study. Benchmark.: Int. J., 13: 135-146.

72. Nahm, A.Y., M.A. Vonderembse, S.S. Rao and T.S. Ragu-Nathan, 2006. Time-based manufacturing improves business performance-results from a survey. Int. J. Prod. Econ., 101: 213-229.

73. Narain, R., R.C. Yadav and J. Sarkis, 2007. Investment justification of advanced manufacturing technology: A review. Int. J. Services Operat. Manage., 3: 41-73.

74. Nemetz, P.L. and L.W. Fry, 1988. Flexible manufacturing organizations: Implications for strategy formulation and organization design. Acad. Manage. Rev., 13: 627-638.

75. Noori, H., 1997. Implementing advanced manufacturing technology: The perspective of a newly industrialized country (Malaysia). J. High Technol. Manage. Res., 8: 1-20.

76. Park, Y.T., 2000. National systems of Advanced Manufacturing Technology (AMT): Hierarchical classification scheme and policy formulation process. Technovation, 20: 151-159.

77. Parthasarthy, R. and S.P. Sethi, 1992. The impact of flexible automation on business strategy and organizational structure. Acad. Manage. Rev., 17: 86-111.

78. Preece, D., 1995. Organisation and Technical Change. Routledge, London.

79. Ratnasingam, J., K. Wagner and S.R. Albakshi, 2009. The impact of iso 14001 on production management practices: A survey of Malaysian wooden furniture manufacturers. J. Applied Sci., 9: 4081-4085.

80. Rosnah, M.Y., M.M.H.M. Ahmad, S. Sulaiman and Z. Mohamad, 2003. Increasing competitiveness through advanced manufacturing technologies. Int. J. Manuf. Technol. Manage., 5: 371-379.

81. SMIDEC, 2007. SME Handbook, Policies, Incentives, Programmes and Financial Assistance for SMEs. SMIDEC, Kula Lumpur, Malaysia.

82. Sambasivarao, K.V. and S.G. Deshmukh, 1995. Selection and implementation of advanced manufacturing technologies classification and literature review of issues. Int. J. Operat. Prod. Manage., 15: 43-62.

83. Sanchez, A.M., 1996. Adopting advanced manufacturing technologies: Experience from Spain. J. Manuf. Syst., 15: 133-140.

84. Schroder, R. and A. S. Sohal, 1999. Organisational characteristics associated with AMT adoption: Towards a contingency framework. Int. J. Operat. Prod. Manage., 19: 1270-1291.

85. Shepherd, D.A., C. mcdermott and G.N. Stock, 2000. Advanced manufacturing technology: Does more radicalness mean more perceived benefits?. J. High Technol. Manage. Res., 11: 19-33.

86. Siegel, D.S., D.A. Waldman and W.E. Youngdahl, 1997. The adoption of advanced manufacturing technologies: Human resource management implications. IEEE Trans. Eng. Manage., 44: 288-298.

87. Small, M.H. and I.J. Chen, 1995. Investment justification of advanced manufacturing technology: An empirical analysis. J. Eng. Technol. Manage., 12: 27-55.

88. Small, M.H. and I.J. Chen, 1997. Economic and strategic justification of AMT inferences from industrial practices. Int. J. Prod. Econ., 49: 65-75.

89. Small, M.H. and M. Yasin, 2000. Human factors in the adoption and performance of advanced manufacturing technology in unionized firms. Ind. Manage. Data Syst., 100: 389-402.

90. Small, M.H. and M.M. Yasin, 1997. Advanced manufacturing technology: Implementation policy and performance. J. Operat. Manage., 15: 349-370.

91. Small, M.H. and M.M. Yasin, 1997. Developing a framework for the effective planning and implementation of advanced manufacturing technology. Int. J. Operat. Prod. Manage., 17: 468-489.

92. Small, M.H., 1993. Towards successful implementation of advanced manufacturing technology: A process-factors-process approach. DBA Thesis, Cleveland State University, pp: 351.

93. Small, M.H., 2006. Justifying investment in advanced manufacturing technology: A portfolio analysis. Ind. Manage. Data Syst., 106: 485-508.

94. Small, M.H., 2007. Planning, justifying and installing advanced manufacturing technology: A managerial framework. J. Manuf. Technol. Manage., 18: 513-537.

95. Song, J.B., D.S. Dai and Y.Q. Song, 2007. The relationship between change of organizational structure and implementation of advanced manufacturing technology: An empirical study. Proceedings of the International Conference on Management Science and Engineering.

96. Stock, G.N. and C.M. McDermott, 2001. Organizational and strategic predictors of manufacturing technology implementation success: An exploratory study. Technovation, 21: 625-636.

97. Stuart, O., 2002. A comparison of AMT strategies in the USA, South Africa and Germany. Int. J. Manuf. Technol. Manage., 4: 441-453.

98. Sun, H. and F. Gertsen, 1995. Organizational changes related to advanced manufacturing technology in the production area. Int. J. Prod. Econ., 41: 369-375.

99. Sun, X.L., Y.Z. Tian and G.G. Cui, 2007. The empirical study on the impact of advanced manufacturing technology on organizational structure and human resources management. Proceedings of the 14th International Conference on Management Science and Engineering, Aug. 20-22, IEEE Xplore, London, pp: 1548-1553.

100. Suresh, N. and J.R. Meredith, 1985. Justifying multi machine systems: An integrated strategic approach. J. Manuf. Syst., 4: 117-134.

101. Swamidass, P. and W. Newell, 1987. Manufacturing strategy, environmental uncertainty and performance: A path analytic model. Manage. Sci., 33: 509-524.

102. Swamidass, P.M. and S. Kotha, 1998. Explaining manufacturing technology use, firm size and performance using a multidimensional view of technology. J. Operat. Manage., 17: 23-37.

103. Vickery, G. and E. Blau, 1989. Gouemment Policies and the Diffusion of Microelectronics. Organisation for Economic Co-operation and Development, Paris.

104. Waldeck, N.E. and Z.M. Leffakis, 2007. HR perceptions and the provision of workforce training in an AMT environment: An empirical study. Omega, 35: 161-172.

105. Waldeck, N.E., 2007. Worker assessment and the provision of developmental activities with advanced technology: An empirical study. Int. J. Prod. Econ., 107: 540-554.

106. Youssef, M.A., 1992. Getting to know advanced manufacturing technologies. Ind. Eng., 24: 40-42.

107. Yusuff, R.M., M.M.H.M. Ahmad and M.R. Osman, 2004. Barriers to AMT implementation in the SMIs of a developing country. Int. J. Eng. Technol., 1: 39-46.

108. Yusuff, R.M., M.M.H.M. Ahmad, S. Sulaiman and Z. Mohamad, 2004. Organization adaptation for AMT implementation in the SMIs. Int. J. Eng. Technol., 1: 131-138.

109. Yusuff, R.M., M.R. Osman and M.S.J. Hashmi, 1997. A simulation study of the applications of JIT and cellular manufacturing concepts in SMI. Int. J. Flexible Automation Integrated Manuf., 5: 105-122.

110. Yusuff, R.M., M.S.J. Hashmi and L.W. Chek, 2005. Advanced manufacturing technologies in SMEs: Strategic requirements for implementation in a developing country. Asia Pacific Technol. Monitoring J., 22: 23-29.

111. Yusuff, R.M., S. Saberi and N. Zulkifli, 2008. A comparison on the capabilities of malaysian SMEs with different equity structure in implementing advanced manufacturing technologies. AIJSTPME., 1: 63-75.

112. Zammuto, R.F. and E.J. O'Connor, 1992. Gaining advanced manufacturing technologies benefits: The roles of organization design and culture. Acad. Manage. Rev., 17: 701-728.

113. Zhang, Q., M.A. Vonderembse and M. Cao, 2006. Achieving flexible manufacturing competence the roles of advanced manufacturing technology and operations improvement practices. Int. J. Operat. Prod. Manage., 26: 580-599.

114. Zhao, B., A. Verma and B. Kapp, 1992. Implementing advanced manufacturing technology in organizations: A socio-technical systems analysis. Proceedings of the IEEE International Engineering Management Conference.

115. Zhao, H. and H.C. Co, 1997. Adoption and implementation of advanced manufacturing technology in Singapore. Int. J. Prod. Econ., 48: 7-19.

Chapter 3

EFFECTS OF COMPUTER-AIDED MANUFACTURING TECHNOLOGY ON PRECISION OF CLINICAL METAL-FREE RESTORATIONS

Ki-Hong Lee[1], In-Sung Yeo[1], Benjamin M. Wu[2], Jae-Ho Yang[1], Jung-Suk Han[1], Sung-Hun Kim[1], Yang-Jin Yi[1], and Taek-Ka Kwon[3]

[1]Department of Prosthodontics, School of Dentistry and Dental Research Institute, Seoul National University, Daehak-ro, Jongno-gu, Seoul 110-749, Republic of Korea

[2]Division of Advanced Prosthodontics, UCLA School of Dentistry, Los Angeles, CA 90095, USA

[3]Department of Dentistry, St. Vincent Hospital, Catholic University of Korea, Ji-dong, Paldal-gu, Suwon 442-723, Republic of Korea

ABSTRACT

Purpose. The purpose of this study was to investigate the marginal fit of metal-free crowns made by three different computer-aided design/computer-aided manufacturing (CAD/CAM) systems. Materials and Methods. The maxillary left first premolar of a dentiform was prepared for all-ceramic crown restoration. Thirty all-ceramic premolar crowns were made, ten each manufactured by the Lava system, Cercon, and Cerec. Ten metal ceramic gold (MCG) crowns served as control. The marginal gap of each sample was measured under a stereoscopic microscope at 75x magnification after cementation. One-way ANOVA and the Duncan's post hoc test were used for data analysis at the significance level of 0.05. Results. The mean (standard deviation) marginal gaps were 70.5 (34.4) μm for the MCG crowns, 87.2 (22.8) μm for Lava, 58.5 (17.6) μm for Cercon, and 72.3 (30.8) μm for Cerec. There were no significant differences in the marginal fit among the groups except that the Cercon crowns had significantly smaller marginal gaps than the Lava crowns . Conclusions. Within the limitation of this study, all the metal-free restorations made by the digital CAD/CAM systems had clinically acceptable marginal accuracy.

INTRODUCTION

With increasing demand for aesthetics, many studies on zirconia, which is the most representative element for metal-free restoration in the field of restorative dentistry, have been recently performed due to its acceptable aesthetics and high strength that is comparable with the strength of a metal ceramic crown [1– 8]. Yttria-stabilized tetragonal zirconia polycrystal is provided as a block form to secure the maximum strength [6, 7]. A new precise mechanical subtracting process has been introduced instead of the previous adding method including waxing, investing, and casting to fabricate a prosthodontic shape from the block. A computer-aided design/computer-aided manufacturing (CAD/CAM) system has been further developed in dentistry over the last 20 years to handle very precise data acquisition, complex restoration design, complete task processing, and high-end cutting system [9].

One of the most important elements in evaluating a fixed prosthodontic device is marginal accuracy. Every prosthodontic restoration process, from abutment preparation to cementation, has effects on the marginal fit of the restoration [10]. Unlike the traditional analogue methods, the CAD/CAM system needs the precision of the system itself, including the accurate digital conversion of acquired information and calibration of the digitalized data according to materials used in CAM. Therefore, it is important in clinical CAD/CAM application to prosthodontic restoration to understand both the differences between the CAD/CAM systems and the accuracy of the resulting crowns.

This study aimed to investigate the marginal fit of zirconia crowns made by widely used CAD/CAM systems: Lava (3M ESPE, Seefeld, Germany), Cercon (DeguDent, Hanau, Germany), and Cerec (Sirona Dental Systems GmbH, Bensheim, Germany). This study also compared the marginal fit of the zirconia crowns with that of a metal ceramic gold (MCG) crown, which is one of the restoration forms clinically used for the longest period.

MATERIALS AND METHODS

The maxillary left first premolar (#24) of the dentiform (Columbia Dentoform Corp., New York) was prepared to form an abutment tooth. Two millimeters of the occlusal surface and 1.0–1.4 mm of the lateral side were reduced. The completed convergence angles of the abutment were about 8–10° both mesiodistally and buccolingually. The margin was assigned with 1 mm of a heavy chamfer margin in the overall range of the cervical aspect (Figure 1). After the abutment preparation, the resin tooth was invested onto the plaster, and the impression was acquired by using the additional silicone impression products of putty and light body (Exafine, GC Co., Tokyo, Japan). Forty original resin models (Exakto-Form, Bredent, Senden, Germany) were manufactured from the silicone impression. These resin models were subsequently used for the measurements of the marginal openings after the final restorations were cemented to these models. The models were divided into 4 groups by assigning 10 models to each group. Lava, Cercon, and Cerec systems were used to fabricate final restorations. Ten single MCG premolar crowns served as control, which were made by the conventional casting method. The other all-ceramic crowns were fabricated according to the manufacturers' recommendations of the systems evaluated in this study. The gap for cement was all assigned as 60 μm.

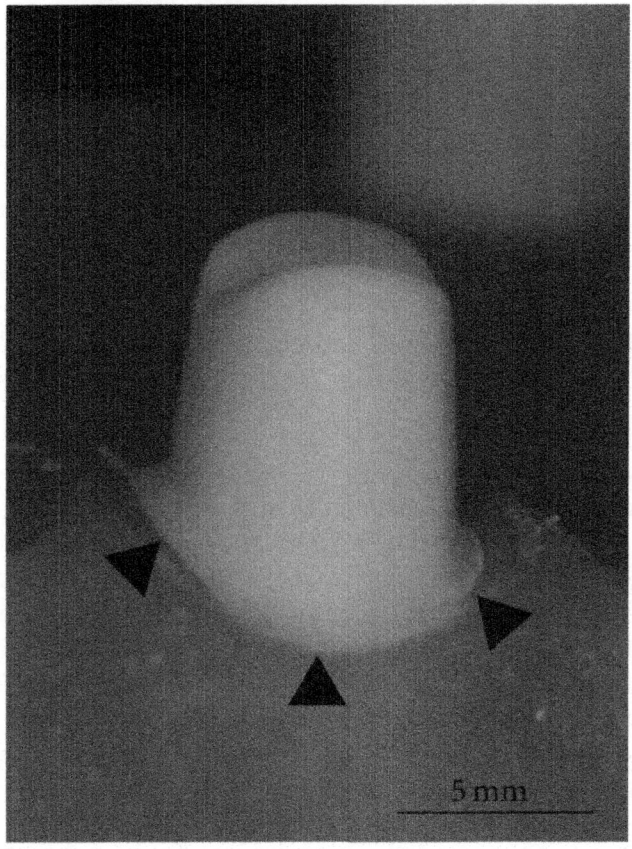

(a)

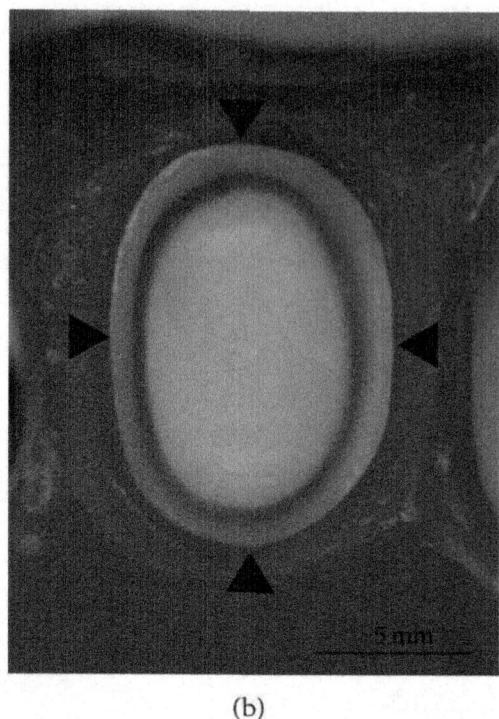

(b)

Figure 1: The prepared premolar resin tooth from the buccal (a) and occlusal viewpoints (b). Note the circumferential cervical margin of 1 mm width (black arrowheads).

The working die productions for the MCG, Lava, and Cercon crowns were performed using high strength dental stone (GC Fujirock EP, GC Europe N.V., Leuven, Belgium) after taking impressions of the original resin models with the additional silicone impression materials of putty and light body (Exafine). The virtual working dies for the Cerec crowns were produced by direct scanning method. For the production of MCG crowns, a wax pattern was produced by using a conventional method with high strength dental stone model. The die spacer (Pico-Fit Die Spacer Varnish (silver), Renfert USA, Il, USA) was coated 3 times on high strength dental stone. Considering the fact that 1 time die spacer coating creates a layer thickness of 14–20 µm from the manufacturer's technical data, the practice allowed for a cement space of approximately 42–60 µm. The gold (Bio Herador SG, Heraeus, Germany) coping was produced by following the investing and casting procedures and then veneered with porcelain. For the Lava crowns, the high strength dental stone dies were scanned with a scanner (Lava Scan Scanner) and zirconia copings were designed under a CAD system (Lava CAD), which gave the cement space of 60 μm. The copings were produced by milling zirconia blocks (Lava zirconia blocks) with a CAM system (Lava Form Milling Unit). The copings were manufactured by setting the thickness of the coping at 0.5 mm. The final crowns were completed by veneering porcelain (Lave Ceram) on the copings after sintering. In the manufacturing of Cercon crowns, the working dies were also scanned using a scanner (Cercon EYE) and zirconia frameworks were designed using a CAD software (Cercon ART). Zirconia blocks (Cercon zirconia blocks) were then milled using a CAM system (Cercon BRAIN), to make the frameworks that were 0.5 mm thick. The milled zirconia frameworks were sintered and were veneered with a heat-pressed material (IPS e.max Ceram, Ivoclar Vivadent AG, Benderer Str. 2, Liechtenstein) and technique, to manufacture the final crowns. For the fabrication of Cerec crowns, the original resin models were directly scanned (CEREC Bluecam) to make the software working dies and the final crowns were designed using a CAD software (CEREC 3D). Zirconia blocks (IPS e.max ZirCAD, Ivoclar Vivadent AG, Schaan, Liechtenstein) were milled by a CAM system (CEREC in Lab MC XL milling machine) and sintered to make the final restorations with no veneering procedure. The procedures, instruments, and materials to make the specimens are summarized in Figure 2.

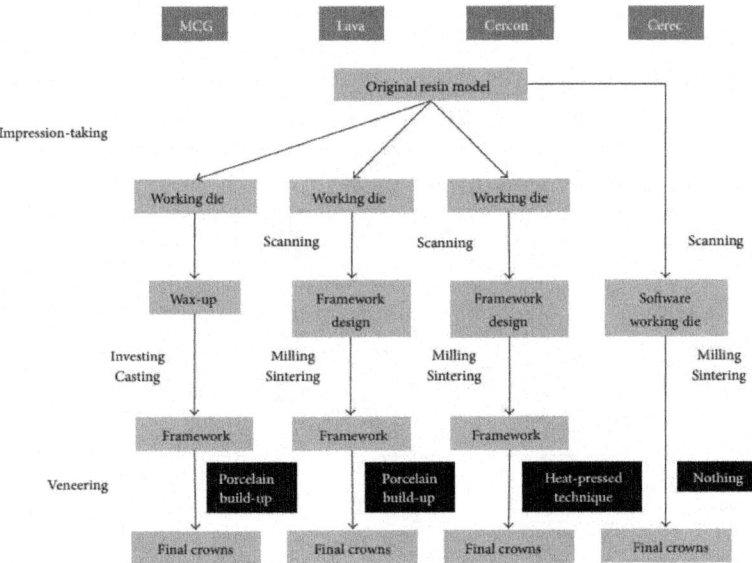

Figure 2: Summarized fabrication procedures for each system. Note that there was neither impression-taking nor veneering procedure in the Cerec system, which manufactured the purely digitalized crowns. Also, notice that each all-ceramic system used a different veneering technique. MCG; metal ceramic gold crowns.

The MCG, Lava, Cercon, and Cerec crowns were, respectively, cemented to their own resin models by using a resin cement (RelyX Unicem Clicker, 3M ESPE, Germany). During cement setting time, 50 N loading was applied with finger pressure by the person who had trained to calibrate the 50 N load with a laboratory scale. Excessive cement material was cleaned with cotton pellets. The marginal fit of each sample was measured by using a stereoscopic microscope (Nikon DS-Fi 1, Nikon, Japan) at 75x magnification. The marginal gap was defined in this study as a distance on the microscope from a point of the tooth margin to the intersecting point between the restoration margin and the line perpendicular to the tangent line to the tooth margin at the tooth margin point. For each crown, the gap was measured at one point of the labial, lingual, mesial, and distal surface. The marginal gap of a crown was calculated as the mean of the measured four gaps.

The mean and standard deviation (SD) were calculated for the measured marginal gaps of each group. One-way ANOVA and a post hoc test, Duncan's test, were used to find any statistically significant difference among the groups at the level of significance of 0.05.

RESULTS

The mean marginal gaps (SD) of MCG, Lava, Cercon, and Cerec crowns were determined to be 70.5 (34.4) μm, 87.2 (22.8) μm, 58.5 (17.6) μm, and 72.3 (30.8) μm, respectively. The descriptive statistics including the mean, SD, minimum, and maximum measured values of each group are presented in Table 1. One-way ANOVA and Duncan's post hoc test showed that there were no significant differences in the marginal fit among the groups except that the Cercon crowns had significantly smaller marginal gaps than the Lava crowns ($P < 0.001$).

Table 1: Mean, SD, and minimum and maximum values of the marginal fit for each of the groups.

Group	Mean (SD)	Minimum–maximum
MCG	70.5 (34.4)^{AB*}	31.9–207.7
Lava	87.2 (22.8)^{B}	45.1–140.9
Cercon	58.5 (17.6)^{A}	34.8–97.3
Cerec	72.3 (30.8)^{AB}	21.8–164.1

* The groups with the same superscript letters (A and B) were not significantly different (unit; μm).

DISCUSSION

There are many and various criteria about the clinically acceptable marginal fit of prosthodontic restoration [11–15]. ADA specification number 8 defined that the range should be 25–40 μm, and Ostlund stated that the value should not exceed 50 μm [11]. Unfortunately, those values appear to be very difficult to obtain clinically. Christensen reported that a maximum marginal distance of 119 μm was allowed by dentists for the proximal surface of gold inlays through observations using eyes, probes, and radiographic images and stated an approximate 39 μm maximum marginal distance for the occlusion surface [12]. McLean and von Fraunhofer stated that a marginal gap of about 100 μm does not cause any clinical problems in a study observing 1,000 dental restorations performed over more than 5 years, concluding that the clinically allowable maximum marginal discrepancy was 120 μm [13]. Another previous study evaluated that a marginal gap up to 100 μm was clinically acceptable, while still another extended the clinically acceptable marginal gap to 200 μm [14, 15]. There is still controversy over the clinically acceptable marginal fit standard. However, most authors are considered to agree upon the fact that the marginal discrepancy should be less than 200 μm [16–23].

The measurement values that were acquired in the present study were in the clinically acceptable range for all the test groups. Most of the currently used CAD/CAM systems were found to show appropriate clinical marginal fit by exhibiting a mean marginal discrepancy value of less than 200 μm. Bindl and Mörmann found no significant difference in the marginal fit of crowns, when comparing the marginal fit of CAD/CAM all-ceramic crowns of Cerec inLab, DCS, Decim, and Procera, the slip cast type crown of In-Ceram zirconia, and heat-pressing type crown of Empress 2, showing a marginal opening range of about 20–70 μm [24]. The marginal fit of the 4-unit fixed dental prostheses made by four CAD/CAM systems (Cercon, Cerec inLab, Digident, Everest) was evaluated to be 57.9–206.3 μm [25]. Another previous study investigating the marginal accuracy of 3-unit fixed dental prostheses showed the mean marginal gaps of 77–92 μm for the Cerec inLab, Digident, and Lava systems [26]. The previous results were similar to those of this study although there were some numerical differences according to the experimental conditions including the restored teeth (anterior, posterior), the restoration types (single, multiple), and the fabrication procedures.

The Cercon premolar crowns exhibited significantly superior marginal fit to the Lava crowns in this study. However, these statistics were unable to be interpreted as superiority of one system in precision to the other because there were no significant differences either between the Cercon and the control (MCG) groups or between Lava and control. Differences in the veneer techniques, not those in the CAD/CAM systems, could explain some causes of the results shown in this study. Some previous studies showed the differences in accuracy between the restorations with and without the porcelain build-up procedures and the significant effects of the veneering methods on restoration precision [27, 28]. This investigation, however, did not consider a CAD/CAM system and a veneer technique as two independent variables, which was one of the limitations. Further studies are required to evaluate and to compare the effects of those two factors, the systems and veneering methods, on the marginal accuracy of prosthodontic restorations. In addition, this study indicated that the accuracy of a dental restoration fabricated by digital technology may be clinically acceptable, when compared with that by conventional analogue method. However, various approaches were found according to the CAD/CAM systems: pure digital techniques and digital-analogue combinations, as shown in Figure 2. Further studies are needed to compare each step in digital procedures with that in analogue.

CONCLUSIONS

Computer-aided digital technologies may manufacture metal-free restorations that are clinically acceptable in precision. Considering the results in this study, the marginal gaps of the digitalized metal-free crowns were similar to those of the conventional metal ceramic gold crowns. All the accuracy investigated in this study may be within the generally agreed clinically acceptable marginal fit standard.

ACKNOWLEDGMENT

This work was supported by the Overseas Research Program of Seoul National University Dental Hospital, Seoul, Korea.

REFERENCES

1. F. Beuer, H. Aggstaller, D. Edelhoff, W. Gernet, and J. Sorensen, "Marginal and internal fits of fixed dental prostheses zirconia retainers," Dental Materials, vol. 25, no. 1, pp. 94–102, 2009.

2. R. C. de Oyagüe, M. I. Sánchez-Jorge, A. S. Turrión, F. Monticelli, M. Toledano, and R. Osorio, "Influence of CAM vs. CAD/CAM scanning methods and finish line of tooth preparation in the vertical misfit of zirconia bridge structures," American Journal of Dentistry, vol. 22, no. 2, pp. 79–83, 2009.

3. A. N. Cavalcanti, R. M. Foxton, T. F. Watson, M. T. Oliveira, M. Giannini, and G. M. Marchi, "Y-TZP ceramics: key concepts for clinical application," Operative Dentistry, vol. 34, no. 3, pp. 344–351, 2009.

4. V. Covacci, N. Bruzzese, G. Maccauro et al., "In vitro evaluation of the mutagenic and carcinogenic power of high purity zirconia ceramic," Biomaterials, vol. 20, no. 4, pp. 371–376, 1999.

5. S. O. Koutayas, T. Vagkopoulou, S. Pelekanos, P. Koidis, and J. R. Strub, "Zirconia in dentistry. Part 2. Evidence-based clinical breakthrough," The European Journal of Esthetic Dentistry, vol. 4, no. 4, pp. 348–380, 2009.

6. R. G. Luthardt, O. Sandkuhl, and B. Reitz, "Zirconia-TZP and alumina— advanced technologies for the manufacturing of single crowns," The European Journal of Prosthodontics and Restorative Dentistry, vol. 7, no. 4, pp. 113–119, 1999. View at Google Scholar · View at Scopus

7. C. Piconi and G. Maccauro, "Zirconia as a ceramic biomaterial," Biomaterials, vol. 20, no. 1, pp. 1–25, 1999.

8. T. Vagkopoulou, S. O. Koutayas, P. Koidis, and J. R. Strub, "Zirconia in dentistry: part 1. Discovering the nature of an upcoming bioceramic," The European Journal of Esthetic Dentistry, vol. 4, no. 2, pp. 130–151, 2009.

9. T. Miyazaki, Y. Hotta, J. Kunii, S. Kuriyama, and Y. Tamaki, "A review of dental CAD/CAM: current status and future perspectives from 20 years of experience," Dental Materials Journal, vol. 28, no. 1, pp. 44–56, 2009.

10. T. F. Alghazzawi, P.-R. Liu, and M. E. Essig, "The effect of different fabrication steps on the marginal adaptation of two types of glass-infiltrated ceramic crown copings fabricated by CAD/CAM technology," Journal of Prosthodontics, vol. 21, no. 3, pp. 167–172, 2012.

11. L. E. Ostlund, "Cavity design and mathematics: their effect on gaps at the margins of cast restorations,"Operative Dentistry, vol. 10, no. 4, pp. 122–137, 1985.

12. G. J. Christensen, "Marginal fit of gold inlay castings," The Journal of Prosthetic Dentistry, vol. 16, no. 2, pp. 297–305, 1966.

13. J. W. McLean and J. A. von Fraunhofer, "The estimation of cement film thickness by an in vivo technique," British Dental Journal, vol. 131, no. 3, pp. 107–111, 1971.

14. F. Sulaiman, J. Chai, L. M. Jameson, and W. T. Wozniak, "A comparison of the marginal fit of In-Ceram, IPS Empress, and Procera crowns," The International Journal of Prosthodontics, vol. 10, no. 5, pp. 478–484, 1997.

15. I. Gulker, "Margins," The New York State Dental Journal, vol. 51, no. 4, pp. 213–217, 1985.

16. S. M. Beschnidt and J. R. Strub, "Evaluation of the marginal accuracy of different all-ceramic crown systems after simulation in the artificial mouth," Journal of Oral Rehabilitation, vol. 26, no. 7, pp. 582–593, 1999.

17. F. Beuer, M. Naumann, W. Gernet, and J. A. Sorensen, "Precision of fit: zirconia three-unit fixed dental prostheses," Clinical Oral Investigations, vol. 13, no. 3, pp. 343–349, 2009.

18. M. Kern, H. G. Schaller, and J. R. Strub, "Marginal fit of restorations before and after cementation in vivo," The International Journal of Prosthodontics, vol. 6, no. 6, pp. 585–591, 1993.

19. J. Kunii, Y. Hotta, Y. Tamaki et al., "Effect of sintering on the marginal and internal fit of CAD/CAM-fabricated zirconia frameworks," Dental Materials Journal, vol. 26, no. 6, pp. 820–826, 2007.

20. P. Pera, S. Gilodi, F. Bassi, and S. Carossa, "In vitro marginal adaptation of alumina porcelain ceramic crowns," The Journal of Prosthetic Dentistry, vol. 72, no. 6, pp. 585–590, 1994.

21. A. J. Raigrodski, "Contemporary materials and technologies for all-ceramic fixed partial dentures: a review of the literature," The Journal of Prosthetic Dentistry, vol. 92, no. 6, pp. 557–562, 2004.

22. P. Vigolo and F. Fonzi, "An in vitro evaluation of fit of zirconium-oxide-based ceramic four-unit fixed partial dentures, generated with three different CAD/CAM systems, before and after porcelain firing cycles and after glaze cycles," Journal of Prosthodontics, vol. 17, no. 8, pp. 621–626, 2008.

23. I.-S. Yeo, J.-H. Yang, and J.-B. Lee, "In vitro marginal fit of three all-ceramic crown systems," The Journal of Prosthetic Dentistry, vol. 90, no. 5, pp. 459–464, 2003.

24. A. Bindl and W. H. Mörmann, "Marginal and internal fit of all-ceramic CAD/CAM crown-copings on chamfer preparations," Journal of Oral Rehabilitation, vol. 32, no. 6, pp. 441–447, 2005.

25. P. Kohorst, H. Brinkmann, J. Li, L. Borchers, and M. Stiesch, "Marginal accuracy of four-unit zirconia fixed dental prostheses fabricated using different computer-aided design/computer-aided manufacturing systems," European Journal of Oral Sciences, vol. 117, no. 3, pp. 319–325, 2009.

26. S. Reich, M. Wichmann, E. Nkenke, and P. Proeschel, "Clinical fit of all-ceramic three-unit fixed partial dentures, generated with three different

CAD/CAM systems," European Journal of Oral Sciences, vol. 113, no. 2, pp. 174–179, 2005.

27. M. Vojdani, A. Safari, M. Mohaghegh, et al., "The effect of porcelain firing and type of finish line one the marginal fit of zirconia copings," Journal of Dentistry, vol. 16, no. 2, pp. 113–120, 2015.

28. K. Torabi, M. Vojdani, R. Giti, M. Taghva, and S. Pardis, "The effect of various veneering techniques on the marginal fit of zirconia copings," The Journal of Advanced Prosthodontics, vol. 7, no. 3, pp. 233–239, 2015.

Chapter 4

A REVIEW OF ADDITIVE MANUFACTURING

Kaufui V. Wong and Aldo Hernandez

Department of Mechanical and Aerospace Engineering, University of Miami, Coral Gables, FL 33146, USA

ABSTRACT

Additive manufacturing processes take the information from a computer-aided design (CAD) file that is later converted to a stereolithography (STL) file. In this process, the drawing made in the CAD software is approximated by triangles and sliced containing the information of each layer that is going to be printed. There is a discussion of the relevant additive manufacturing processes and their applications. The aerospace industry employs them because of the possibility of manufacturing lighter structures to reduce weight. Additive manufacturing is transforming the practice of medicine and making work easier for architects. In 2004, the Society of Manufacturing Engineers did a classification of the various technologies and there are at least four additional significant technologies in 2012. Studies are reviewed which were about the strength of products made in additive manufacturing processes. However, there is still a lot of work and research to be accomplished before additive manufacturing technologies become standard in the manufacturing industry because not every commonly used manufacturing material can be handled. The accuracy needs improvement to eliminate the necessity of a finishing process. The continuous and increasing growth experienced since the early days and the successful results up to the present time allow for optimism that additive manufacturing has a significant place in the future of manufacturing.

RAPID PROTOTYPING

The first form of creating layer by layer a three-dimensional object using computer-aided design (CAD) was rapid prototyping, developed in the 1980's for creating models and prototype parts. This technology was created to help the realization of what engineers have in mind. Rapid prototyping is one of the earlier additive manufacturing (AM) processes. It allows for the creation of printed parts, not just models. Among the major advances that this process presented to product development are the time and cost reduction, human interaction, and consequently the product development cycle [1], also the possibility to create almost any shape that could be very difficult to machine. However, at the present time it is not yet adopted in the manufacturing sector, but scientists, medical doctors, students and professors, market researchers, and artists use it [2–4]. With rapid prototyping, scientists and students can rapidly build and analyze models for theoretical comprehension and studies. Doctors can build a model of a damaged body to analyze it and plan better the procedure, market researchers can see what people think of a particular new product, and rapid prototyping makes it easier for artists to explore their creativity.

The steps involved in product development using rapid prototyping are shown in Figure 1. Here, it can be seen that creating models faster save a lot of time and there is the possibility of testing more models.

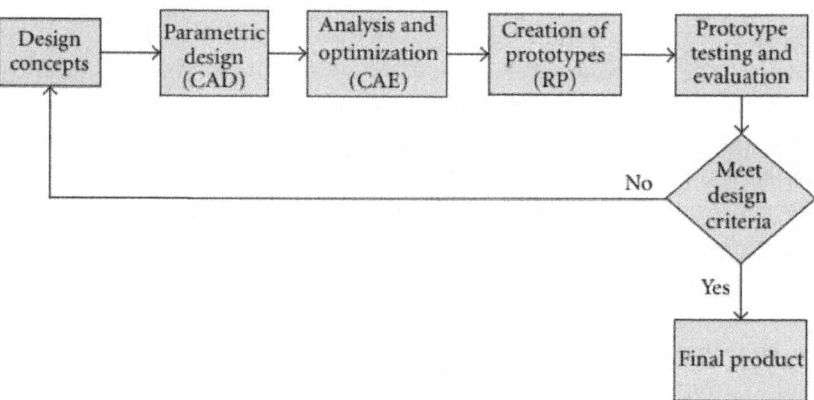

Figure 1: Product development cycle [2].

At the present time, the technologies of rapid prototyping are not just used for creating models, with the advantages in plastic materials it has been possible to create finished products, of course at the beginning they were developed to expand the situations tested in the prototyping process [5]. Nowadays, these technologies have other names like 3D printing, and so forth, but they all have the origins of rapid prototyping [2, 6]. According to Wohler's report 2011 the growth rate for 2010 was 24.1%. The compound annual growth rate for the industry's history, until 2010, is 26.2 percent [7, 8]. This growth shown in Figure 2.

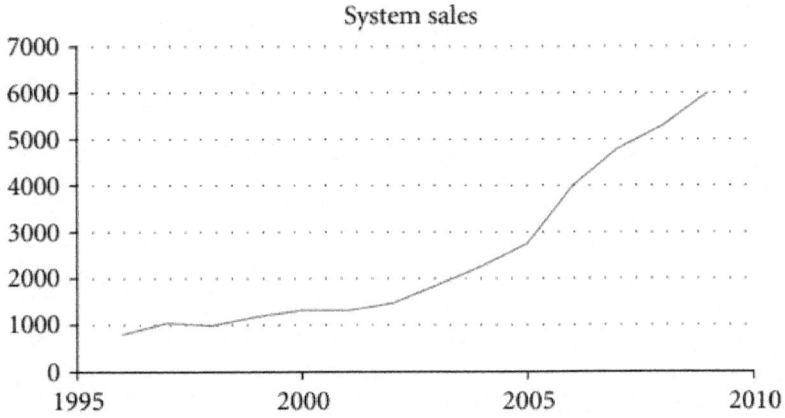

Figure 2: Growth of rapid prototyping [9]. Source. Wohler's report 2010 adapted from [10].

In addition, it is important to notice that rapid manufacturing became possible by other technologies, which are computer-aided design (CAD), computer-aided manufacturing (CAM), and computer numerical control (CNC). This three technologies combined together made possible the printing of three-dimensional objects [2,5, 11].

Rapid prototyping is still not the best solution for all cases, in some cases CNC machining processes still need to be used. Parts dimension could be larger than available additive manufacturing printers [8]. Materials for rapid prototyping are still limited. It is clear that at least it is possible to print metals and ceramics but not all commonly used manufacturing materials [11].

In Figure 3, there is an overview of the different additive manufacturing processes that are going to be further discussed. Here in this figure adapted from [11], the criterion used is to classify these processes into liquid base, solid based, and powder based. The processes included in this review are considered the most relevant in the past, and promising for the future of the industry. The processes considered are stereolithography (SL), Polyjet, fused deposition modeling (FDM), laminated object manufacturing (LOM), 3D printing (3DP), Prometal, selective laser sintering (SLS), laminated engineered net shaping (LENS), and electron beam melting (EBM). The liquid-and powder-based processes seem more promising than solid-based processes of which LOM is the predominant one today. In 2004 [11], EBM, Prometal, LENS, and Polyjet were nonexistent.

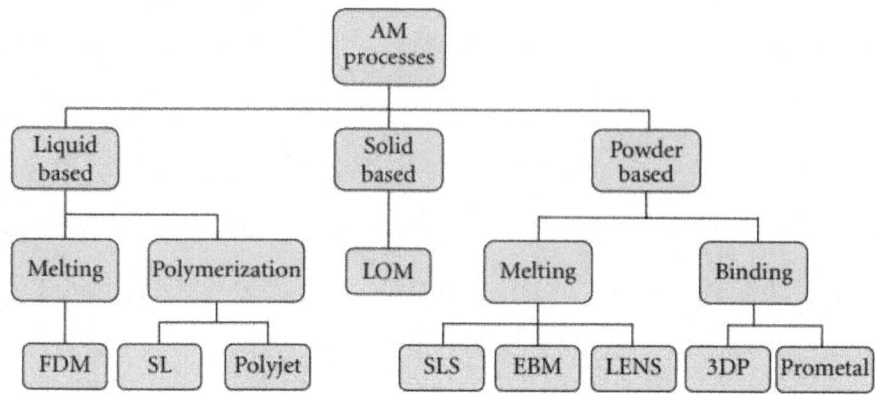

Figure 3: Three-dimensional printing processes. Adapted from [11].

These technologies were first created to produce models, but they have expanded since then. In the chart is presented a survey made by Wohlers in which 24 manufacturers participated and so did 65 services of 5000+ users and costumers. In Figure 4 is shown the amount of responses received by these companies [12].

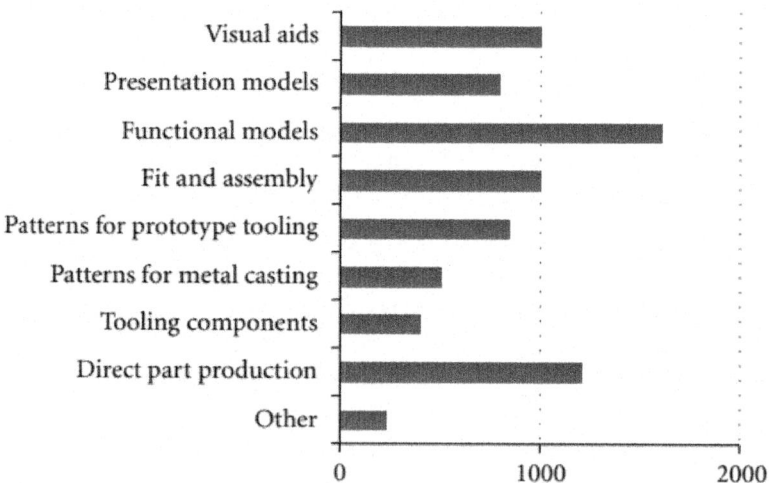

Figure 4: Different usage for additive manufacturing printing processes [12].

STEREOLITHOGRAPHY

Stereolithography (SL), developed by 3D Systems, Inc., was the first and is most widely used process of rapid prototyping, so, in the past the two terms were used synonymously. This is a liquid-based process that consists in the curing or solidification of a photosensitive polymer when an ultraviolet laser makes contact with the resin. The process starts with a model in a CAD software and then it is translated to a STL file in which the pieces are "cut in slices" containing the information for each layer. The thickness of each layer as well as the resolution depend on the equipment used. A platform is built to anchor the piece and supporting the overhanging structures. Then the UV laser is applied to the resin solidifying specific locations of each layer. When the layer is finished the platform is lowered and finally when the process is done the excess is drained and can be reused [2, 5, 11]. A newer version of this process has been developed with a higher resolution and is called microstereolithography. This process that has a layer thickness of less than 10 μm can be achieved [13]. In Figure 5 are shown the basic parts of a stereolithography machine.

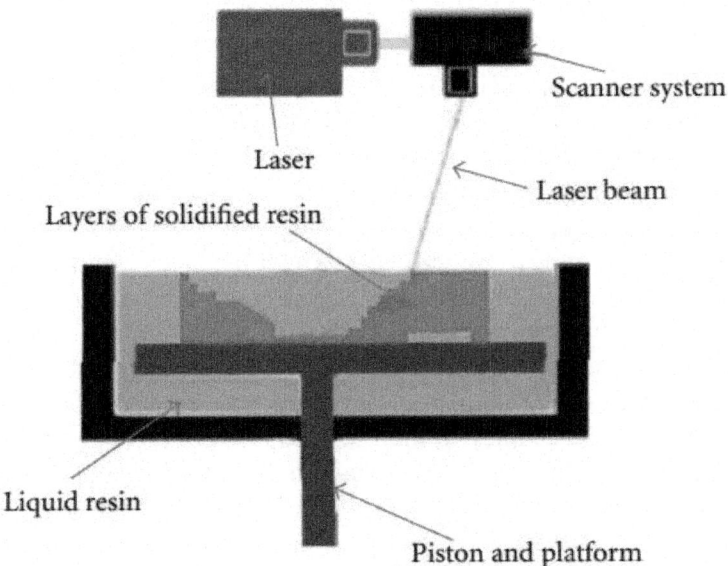

Figure 5: Stereolithography.

The basic principle of this process is the photopolymerization, which is the process where a liquid monomer or a polymer converts into a solidified polymer by applying ultraviolet light which acts as a catalyst for the reactions;

this process is also called ultraviolet curing. It is also possible to have powders suspended in the liquid like ceramics [14].

There are errors induced to the final piece from the process of stereolithography. One is overcuring, which occurs to overhang parts because there is no fusing with a bottom layer. Another is the scanned line shape, which is introduced by the scanning process. Because the resin is a high-viscosity liquid the layer thickness is variable and this introduces an error in the border position control. Another error caused could be if the part needed to have a surface finished process that is normally done by hand [21]. All these errors are minimized in equipments of high quality.

There is the possibility of using different materials while building a piece; this process is called multiple material stereolithography. In order to print with different materials all the resin has to be drained and filled with the new material when the process reach the layer where the change is going to take place. This must occuer even if the first material is going to be used again because is only possible to print consecutive layers. resin. In the software a scheduling process has to be specified [22].

THE STL FILE

The STL file was created in 1987 by 3D Systems Inc. when they first developed the stereolithography, and the STL file stands for this term. It is also called Standard Tessellation Language. There are other types of files, but the STL file is the standard for every additive manufacturing process. The STL file creation process mainly converts the continuous geometry in the CAD file into a header, small triangles, or coordinates triplet list of x, y, and z coordinates and the normal vector to the triangles. This process is inaccurate and the smaller the triangles the closer to reality [2, 13, 23]. The interior and exterior surfaces are identified using the right-hand rule and vertices cannot share a point with a line. Additional edges are added when the figure is sliced. The slicing process also introduces inaccuracy to the file because here the algorithm replaces the continuous contour with discrete stair steps [23]. To reduce this inaccuracy, the technique for a feature that has a small radius in relation to the dimension of the part is to create STL files separately and to combine them later. The dimension in z direction should be designed to have a multiple of the layer thickness value [21]. In Figure 6 is shown the position of the STL file creation in the data flow of a rapid prototyping process. In Figure 7 is shown the data flow in the STL file creation software.

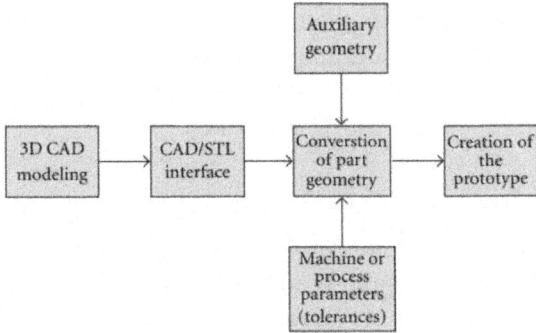

Figure 6: Data flow in rapid prototyping [2].

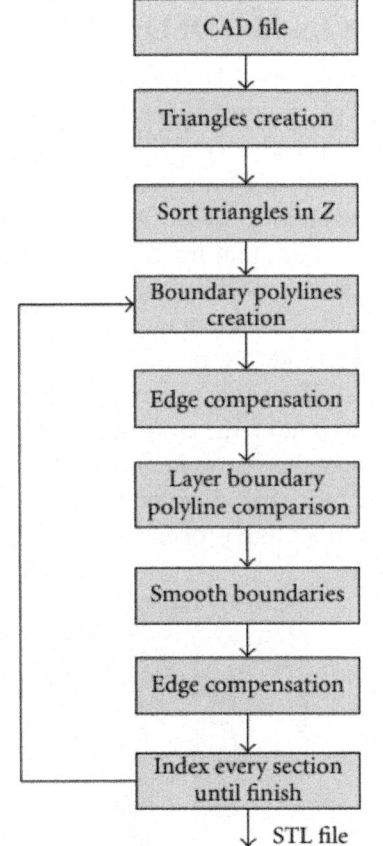

Figure 7: Data flow in STL file creation [2].

Other types of files are stereolithography contour (SLC) and SLI from 3D Systems, CLI from EOS, Hewlett-Packard graphics language (HPGL) from Hewlett-Packard, stereolithography contour from Stratasys, and F&S from Fockele and Schwarze and initial graphics exchange specifications (IGES) [13].

3DP

3DP process is a MIT-licensed process in which water-based liquid binder is supplied in a jet onto a starch-based powder to print the data from a CAD drawing. The powder particles lie in a powder bed and they are glued together when the binder is jetted. This process is called 3DP because of the similarity with the inkjet printing process that is used for two-dimensional printing in paper. This process can handle a high variety of polymers [5, 13].

FUSED DEPOSITION MODELING

Fused deposition modeling (FDM) is an additive manufacturing process in which a thin filament of plastic feeds a machine where a print head melts it and extrude it in a thickness typically of 0.25 mm. Materials used in this process are polycarbonate (PC), acrylonitrile butadiene styrene (ABS), polyphenylsulfone (PPSF), PC-ABS blends, and PC-ISO, which is a medical grade PC. The main advantages of this process are that no chemical post-processing required, no resins to cure, less expensive machine, and materials resulting in a more cost effective process [2, 5]. The disadvantages are that the resolution on the z axis is low compared to other additive manufacturing process (0.25 mm), so if a smooth surface is needed a finishing process is required and it is a slow process sometimes taking days to build large complex parts. To save time some models permit two modes; a fully dense mode and a sparse mode that save time but obviously reducing the mechanical properties [24]. In Figure 8 is shown the basics fused deposition modeling process.

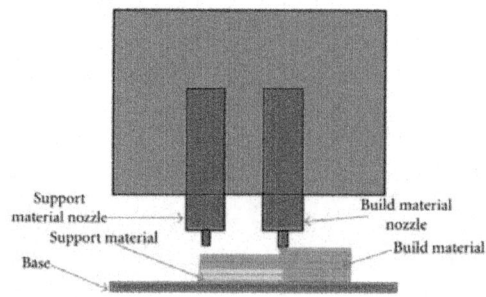

Figure 8: Fused deposition modeling.

PROMETAL

Prometal is a three-dimensional printing process to build injection tools and dies. This is a powder-based process in which stainless steel is used. The printing process occurs when a liquid binder is spurt out in jets to steel powder. The powder is located in a powder bed that is controlled by build pistons that lowers the bed when each layer is finished and a feed piston that supply the material for each layer. After finishing, the residual powder must be removed. When building a mold no postprocessing is required. If a functional part is being built, sintering, infiltration, and finishing processes are required [5, 11]. In the sintering process, the part is heated to 350°F for 24 hour hardening the binder fusing with the steel in a 60% porous specimen. In the infiltration process, the piece is infused with bronze powder when they are heated together to more than 2000°F in an alloy of 60% stainless steel and 40% bronze [25–27]. The same process, but with different sintering temperatures and times, has been used with other materials like a tungsten carbide powder sintered with a zirconium copper alloy for the manufacturing of rocket nozzles; these parts have better properties than CNC machined parts of the same material [28].

SELECTIVE LASER SINTERING

This is a three-dimensional printing process in which a powder is sintered or fuses by the application of a carbon dioxide laser beam. The chamber is heated to almost the melting point of the material. The laser fused the powder at a specific location for each layer specified by the design. The particles lie loosely in a bed, which is controlled by a piston, that is lowered the same amount of the layer thickness each time a layer is finished. This process offers a great variety of materials that could be used: plastics, metals, combination of metals, combinations of metals and polymers, and combinations of metals and ceramics [13, 29, 30]. Examples of the polymers that could be used are acrylic styrene and polyamide (nylon), which show almost the same mechanical properties as the injected parts [29, 31]. It is also possible to use composites or reinforced polymers, that is, polyamide with fiberglass. They also could be reinforced with metals like copper. For metals, a binder is necessary. This could be a polymer binder, which will be later removed by heating or a mix of metals with very different melting point [29, 31, 32]. Parts of alumina with high strength can be built with polyvinyl alcohol, which is an organic binder [5]. The main advantages of this technology are the wide range of materials that can be used. Unused powder can be recycled. The disadvantages are that the accuracy is limited by the size of particles of the material, oxidation needs to be avoided by executing the process in an inert gas atmosphere and for the process to occur at constant temperature near the melting point. This process is also called direct metal laser sintering.

ELECTRON BEAM MELTING

A process similar to SLS is electron beam melting (EBM). This process is relatively new but is growing rapidly. In this process, what melts the powder is an electron laser beam powered by a high voltage, typically 30 to 60 KV. The process takes place in a high vacuum chamber to avoid oxidation issues because it is intended for building metal parts. Other than this, the process is very similar to SLS. EBM also can process a high variety of prealloyed metals. One of the future uses of this process is the manufacturing in outer space [33, 34], since it is all done in a high vacuum chamber.

LASER ENGINEERED NET SHAPING

In this additive manufacturing process, a part is built by melting metal powder that is injected into a specific location. It becomes molten with the use of a high-powered laser beam. The material solidifies when it is cooled down. The process occurs in a closed chamber with an argon atmosphere. This process permits the use of a high variety of metals and combination of them like stainless steel, nickel-based alloys, titanium-6 aluminium-4 vanadium, tooling steel, copper alloys, and so forth. Alumina can be used too. This process is also used to repair parts that by other processes will be impossible or more expensive to do. One problem in this process could be the residual stresses by uneven heating and cooling processes that can be significant in high-precision processes like turbine blades repair [5, 13, 35–37]. Figure 9 is an illustration of how the part is made in this process.

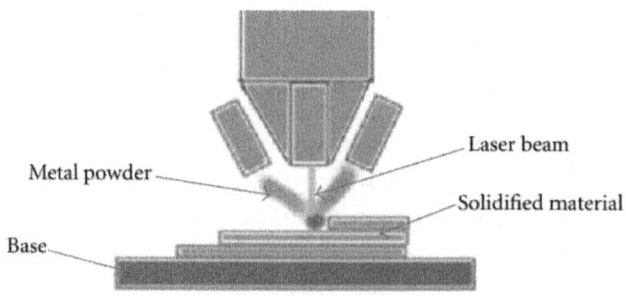

FIGURE 9: Laser engineered net shaping.

Figure 9: Laser engineered net shaping.

LAMINATED OBJECT MANUFACTURING

Laminated Object Manufacturing (LOM) is a process that combines additive and subtractive techniques to build a part layer by layer. In this process the materials come in sheet form. The layers are bonded together by pressure and heat application and using a thermal adhesive coating. A carbon dioxide laser cuts the material to the shape of each layer given the information of the 3D model from the CAD and STL file. The advantages of this process are the low cost, no post processing and supporting structures required, no deformation or phase change during the process, and the possibility of building large parts. The disadvantages are that the fabrication material is subtracted thus wasting it, low surface definition, the material is directional dependent for machinability and mechanical properties, and complex internal cavities are very difficult to be built. This process can be used for models with papers, composites, and metals [2, 5, 38].

POLYJET

This is an additive manufacturing process that uses inkjet technologies to manufacture physical models. The inkjet head moves in the x and y axes depositing a photopolymer which is cured by ultraviolet lamps after each layer is finished. The layer thickness achieved in this process is 16 μm, so the produced parts have a high resolution. However, the parts produced by this process are weaker than others like stereolithography and selective laser sintering. A gel-type polymer is used for supporting the overhang features and after the process is finished this material is water jetted. With this process, parts of multiple colors can be built [9, 39, 40].

APPLICATIONS

Lightweight Machines

With additive manufacturing technologies it is possible to manufacture lightweight parts. In the automotive and aerospace industry the main goal is to make the lightest practical car or aircraft while securing safety. Additive manufacturing technologies have enabled the manufacture of complex cross sectional areas like the honeycomb cell [41] or every other material part that contains cavities and cut-outs which reduce the weight-strength relation. It is possible to create lightweight structures; they are methods to get a shape that have a minimum weight like the hanging method and the soap film method [42]. The hanging method and the soap fill method produce a very difficult form of a structure which has been used for civil construction, but with additive manufacturing it is possible to create structural parts for machines using the shape described by these methods and reducing the total weight. Selective laser sintering, and electron beam are now used in the aircraft and aerospace industries. Engineers perform design within the manufacturing constrains but this process expands the limits. With SLS and EBM, the limit will be the engineer's imagination. They open a whole new dimension of possible designs with almost any prealloyed metal powder [37]. With the traditional process these complex shape structures will be expensive to do if at all possible. With additive manufacturing printing technologies like selective laser sintering or electron beam melting, hollow structures, which are less expensive than a solid one, can be made since less material is used.

Architectural Modeling

Creating an architectural model can be very difficult for architects. Architects usually build their models with hand techniques, but when complex models are on their minds making a physical model can be a very hard task. Modeling is very important for the architects to study the models and their functionality. They are also needed for architects to explain them to their customers and convince them to make the project a reality [43]. Additive manufacturing technologies can provide architects a very powerful tool for their business, by being able to create a physical model faster without worrying about the complexity of their design. It also achieves a better resolution than other processes used in architecture. Architects work with CAD software, so there is no need for them to adapt to anything because the STL file is created from a CAD file. Stereolithography is a process very suitable for the architectural modeling because of the materials used and the printing resolution [34–47].

Medical Applications

Additive manufacturing printing technologies have vast applications in the medical world. They are transforming the practice of medicine through the possibilities of making rapid prototypes and very high quality bone transplants and models of damaged bone of the patients for analysis. Additive manufacturing printing methods permit to scan and build a physical model of defective bones from patients and give doctors a better idea of what to expect and plan better the procedure, this will save cost and time and help achieve a better result [48, 49]. Bone transplants now can be done by printing them and additive manufacturing methods make it possible to have a transplant that is practically identical to the original. Because of the limitless form or shape of what could be built, doctors have the option to create a porous-controlled material that will permit osteoconductivity or to create a precise metal transplant identical to the original depending on the bone to be replaced [50–52]. Characteristics of the transplants such as density, pore shape and size, and pore interconnectivity are important parameters that will manipulate tissue ingrowth and mechanical properties of the implant bone. The mechanical strength of these implants are three to five times higher than others produced by other processes and the possibility of inflammation caused by microdebris that breaks during the procedure is reduced [45]. Additive manufacturing is a very good tool for dentists because they can easily build a plaster model of a patient's mouth or replace the teeth, which have a unique form with process like stereolithography, selective laser sintering and electron beam melting [53–55].

According to PC Magazine [54], an 83-year-old Belgian woman became the first-ever person to receive a transplant jawbone tailor-made for her face using a 3D printer and the surgery time and recovery were a lot less than other patients that received the same procedure. The shapes of bones differ too much between each person and additive manufacturing printing produces transplants that fit better, and are easier to insert and secure, reducing the time for the procedure and produce a better cosmetic result [55, 56].

Stereolithography is being used to manufacture prosthetic sockets. By using this technology to ensure that the form of the socket adapt better to the patient while being more cost-effective than hand or machined methods [57]. Not only hard parts like bones can be produced, also it is possible to print cells in a 3D array that with the possibility of printing complex shapes and arrays human tissue can be printed [58–61]. This technology will help patients that have lost tissue in accidents or from other reasons to recover faster and with better cosmetic results. In addition 3D cell printing technologies offer the possibility of printing artificial blood vessels that can be used in the coronary bypass surgery or any other blood vessel procedure or diseases, like cardiovascular defects and medical therapy [62–64]. The application of this printed blood vessels is in the future. Research in this area, also called bioprinting organs, will eventually lead to printed organs, but this could take 20 years until someone achieves it [65].

Cell printing is not limited to print human tissue; it is also used in the field of molecular electronics. The precision of high-resolution processes like nanolithography and photolithography permits the creation of biochips and biosensors [63, 66, 67].

Improving the Manufacturing of Fuel Cells

Additive manufacturing technologies can be used in processes that require a very precise thin film of a certain material. In the manufacture of polymer electrolyte membrane fuel cells (PEMFCs), it is necessary to precisely deposit a very thin layer of platinum, needed for the oxidation and reduction reactions, with high utilization efficiency of the platinum. This could be critical for taking this technology to the masses. One of the other processes used is screen printing, but this process is done by hand and compromises the uniformity and time efficiency. The process is greatly improved by using the 3D printing process to deposit the layer of platinum. In Figure 10, there is a comparison between the processes. The inkjet printing method is 4 to 5 times faster than screen printing [15].

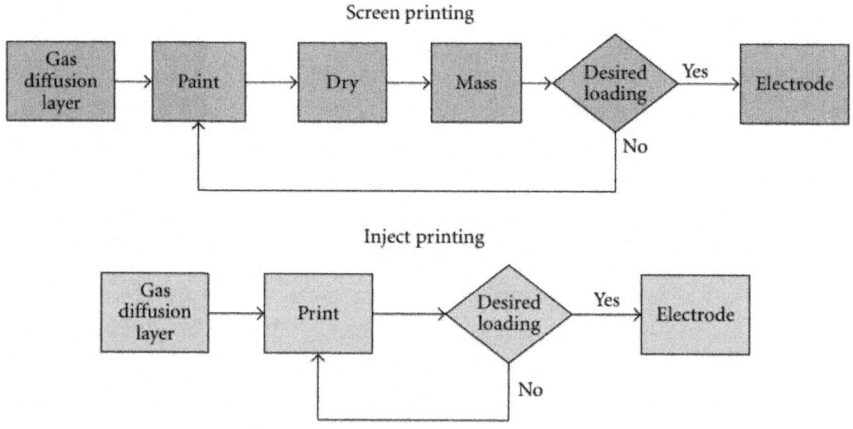

Figure 10: Screen printing compared to inkjet Printing [15].

Additive Manufacturing in Art

Additive manufacturing technologies are a very powerful tool to artist in the fashion, furniture, and lightning industry given the possibility of virtually manufacturing the most complex form imaginable. There are companies that manufacture furnishing complements, lightning, and accessories including clothes using SLS [68–70].

Additive Manufacturing for Hobbyist

Additive manufacturing technologies are reaching nonindustrial users. This revolution started in 2007 with printers that could be as low as $500 using a variant of the fused deposition modeling process. However, these low-cost printers are mostly sold as do-it-yourself (DIY) projects, so, technical ability on the part of the users is required. However, these days with larger companies entering the business 3D printers are closer to reaching the masses. At this time, plastics and edible materials like chocolate materials are only available [71–73]. Users can use CAD software normally used by engineers. Easier ones for hobbyists are available which can be used for design. There are many apps that hobbyists can make 3D models and print them using a 3D printer [74].

Strength Comparison

Studies have been made to analyze the properties of the product in each process. Kim and Oh [16] compared the properties of nonmetal additive manufacturing processes. They tested the specimens in the building direction and perpendicular to the building direction and they found very little influence in the building direction in 3DP but an enormous influence in LOM [70]. Figure 11 is a comparison of the strength for LOM, Polyjet, SL, SLS, FDM, and 3DP.

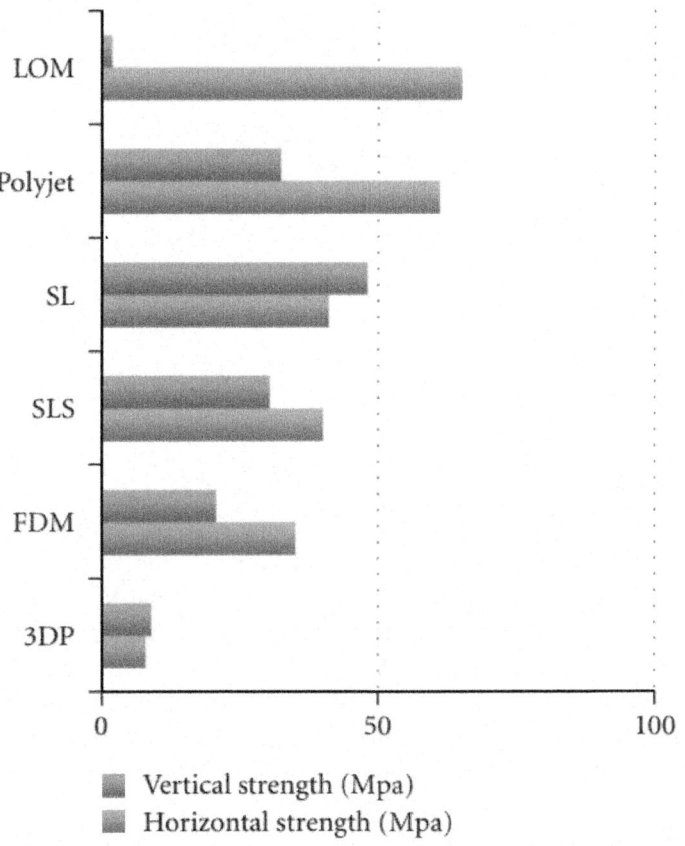

Figure 11: Tensile strength of various 3D printing processes adapted from [16].

The strength of the material produced by Prometal, EBM, SLS, and LENS is compared in Figure 12. The studies were made by [17–20].

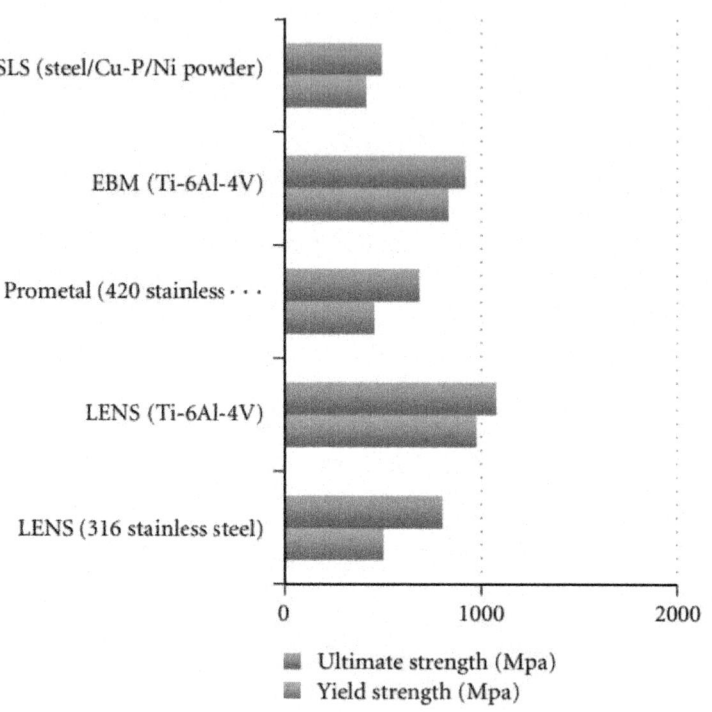

Figure 12: Tensile strength of various 3D printing processes [17–20].

DISCUSSION AND CONCLUSION

In this paper article is discussed the early versions of additive manufacturing for making fast prototypes that was initiated by the necessity of speeding the process in model development and shortening the time between product development and market placement. Additive manufacturing processes take the information from a CAD file that is later converted to an STL file. In this process, the drawing made in the CAD software is approximated by triangles and sliced containing the information of each layer that is going to be printed. There is also a discussion of the relevant additive manufacturing processes and their applications and a review of how the parts are made using these additive manufacturing processes. The continuous and increasing growth experienced since the early days and the successful results up to date, there is optimism that additive manufacturing has a significant place in the future of manufacturing.

In 2004, the Society of Manufacturing Engineers did a classification of the various technologies [11], but there are at least four additional significant technologies in 2012.

Additive manufacturing technologies have been welcomed in the aerospace industry because of the possibility to manufacture lighter structures to reduce weight, which is the common goal of aircraft and spacecraft designers. In the automotive industry, additive manufacturing is advantageous also in reproducing difficult-to-find parts, for example, parts for classic cars. Additive manufacturing is transforming the practice of medicine; now it is possible to have a precise model of a bone before a surgery and the possibility of creating an accurate transplant, no matter how complex its form is. Additive manufacturing is making work easier for architects, who now can print the 3D models of whatever complex shape for a civil project they have in mind. In addition, studies are reviewed which were about the strength of products made in additive manufacturing processes. Review has been presented of studies considering nonmetal material processes and metal material processes in which a comparison of strength of the products was made by the various processes. However, there is still a lot of work and research to be accomplished before additive manufacturing processes become the standard in the manufacturing industry because not every commonly used manufacturing material can be handled. The accuracy needs improvement to eliminate the necessity of a finishing process and to be able to produce parts that require the highest levels of precision.

REFERENCES

1. S. Ashley, "Rapid prototyping systems," Mechanical Engineering, vol. 113, no. 4, p. 34, 1991.

2. R. Noorani, Rapid Prototyping—Principles and Applications, John Wiley & Sons, 2006.

3. J. Flowers and M. Moniz, "Rapid prototyping in technology education," Technology Teacher, vol. 62, no. 3, p. 7, 2002.

4. C. K. Chua, S. M. Chou, S. C. Lin, K. H. Eu, and K. F. Lew, "Rapid prototyping assisted surgery planning," International Journal of Advanced Manufacturing Technology, vol. 14, no. 9, pp. 624–630, 1998.

5. K. Cooper, Rapid Prototyping Technology, Marcel Dekker, 2001.

6. A. Kochan, "Rapid growth for rapid prototyping," Assembly Automation, vol. 17, no. 3, pp. 215–217, 1997.

7. T. Wohlers, Wohlers Report 2011, Wholers Associates, 2011.

8.   T. Wohlers, "Additive Manufacturing Advances," Manufacturing Engineering, vol. 148, no. 4, pp. 55–56, 2012.

9.   T. Wohlers, Wohlers Report 2010, Wholers Associates, 2010.

10.   T. Grimm, User›s Guide to Rapid Prototyping, Society of Manufacturing Engineers, 2004.

11.   P. P. Kruth, "Material incress manufacturing by rapid prototyping techniques," CIRP Annals—Manufacturing Technology, vol. 40, no. 2, pp. 603–614, 1991.

12.   T. Wohlers, Wohlers Report 2009, Wholers Associates, 2009.

13.   J. W. Halloran, V. Tomeckova, S. Gentry et al., "Photopolymerization of powder suspensions for shaping ceramics," Journal of the European Ceramic Society, vol. 31, no. 14, pp. 2613–2619, 2011.

14.   D. T. Pham and C. Ji, "Design for stereolithography," Proceedings of the Institution of Mechanical Engineers, vol. 214, no. 5, pp. 635–640, 2000.

15.   A. D. Taylor, E. Y. Kim, V. P. Humes, J. Kizuka, and L. T. Thompson, "Inkjet printing of carbon supported platinum 3-D catalyst layers for use in fuel cells," Journal of Power Sources, vol. 171, no. 1, pp. 101–106, 2007.

16.   G. D. Kim and Y. T. Oh, "A benchmark study on rapid prototyping processes and machines: quantitative comparisons of mechanical properties, accuracy, roughness, speed, and material cost,"Proceedings of the Institution of Mechanical Engineers, vol. 222, no. 2, pp. 201–215, 2008.

17.   J. P. Kruth, X. Wang, T. Laoui, and L. Froyen, "Lasers and materials in selective laser sintering,"Assembly Automation, vol. 23, no. 4, pp. 357–371, 2003.

18.   L. Facchini, E. Magalini, P. Robotti, and A. Molinari, "Microstructure and mechanical properties of Ti-6Al-4V produced by electron beam melting of pre-alloyed powders," Rapid Prototyping Journal, vol. 15, no. 3, pp. 171–178, 2009.

19.   R. Shivpuri, X. Cheng, K. Agarwal, and S. Babu, "Evaluation of 3D printing for dies in low volume forging of 7075 aluminum helicopter parts," Rapid Prototyping Journal, vol. 11, no. 5, pp. 272–277, 2005.

20.   Y. Xiong, Investigation of the laser engineered net shaping process for nanostructured cermets [ProQuest Dissertations], University of California, 2009.

21.   H. Kim, C. Jae-Won, and R. Wicker, "Scheduling and process planning

for multiple material stereolithography," Rapid Prototyping Journal, vol. 16, no. 4, pp. 232–240, 2010.

22. M. Szilvœi-Nagy and G. Mátyási, "Analysis of STL files," Mathematical and Computer Modelling, vol. 38, no. 7–9, pp. 945–960, 2003.

23. C. Iancu, D. Iancu, and A. Stamcioiu, "From Cad model to 3D print via "STL" file format,"http://www.utgjiu.ro/rev_mec/mecanica/pdf/2010-01/13_Catalin%20Iancu.pdf.

24. S. Morvan, R. Hochsmann, and M. Sakamoto, "ProMetal RCT(TM) process for fabrication of complex sand molds and sand cores," Rapid Prototyping, vol. 11, no. 2, pp. 1–7, 2005.

25. R. C. T. ProMetal, "ProMetal RCT rapid prototyping and digital sand casting services," 2010,http://www.youtube.com/watch?v=Z8MaVaqNr3U.

26. Ex One, "3D metal printing," 2010, http://www.youtube.com/watch?v=i6Px6RSL9Ac&feature=related.

27. D. W. Lipke, Y. Zhang, Y. Liu, B. C. Church, and K. H. Sandhage, "Near net-shape/net-dimension ZrC/W-based composites with complex geometries via rapid prototyping and displacive compensation of porosity," Journal of the European Ceramic Society, vol. 30, no. 11, pp. 2265–2277, 2010.

28. J. P. Kruth, P. Mercelis, J. van Vaerenbergh, L. Froyen, and M. Rombouts, "Binding mechanisms in selective laser sintering and selective laser melting," Rapid Prototyping Journal, vol. 11, no. 1, pp. 26–36, 2005.

29. T. Hwa-Hsing, C. Ming-Lu, and Y. Hsiao-Chuan, "Slurry-based selective laser sintering of polymer-coated ceramic powders to fabricate high strength alumina parts," Journal of the European Ceramic Society, vol. 31, no. 8, pp. 1383–1388, 2011.

30. G. V. Salmoria, R. A. Paggi, A. Lago, and V. E. Beal, "Microstructural and mechanical characterization of PA12/MWCNTs nanocomposite manufactured by selective laser sintering," Polymer Testing, vol. 30, no. 6, pp. 611–615, 2011.

31. D. Slavko and K. Matic, "Selective laser sintering of composite materials technologies," Annals of DAAAM & Proceedings, p. p1527, 2010.

32. Technology Gateway, "NASA| EBF3—electron beam form fabrication," 2009,http://www.youtube.com/watch?v=WrWHwHuWrzk.

33. L. Murr, S. Gaytan, D. Ramirez et al., "Metal fabrication by additive manufacturing using laser and electron beam melting technologies," Journal of Materials Science & Technology, vol. 28, no. 1, pp. 1–14, 2012.

34. C. Semetay, Laser engineered net shaping (LENS) modeling using welding simulation concepts [ProQuest Dissertations and Theses], Lehigh University, 2007.

35. Y. Xiong, Investigation of the laser engineered net shaping process for nanostructured cermets [ProQuest Dissertations and Theses], University of California, 2009.

36. V. K. Balla, S. Bose, and A. Bandyopadhyay, "Processing of bulk alumina ceramics using laser engineered net shaping," International Journal of Applied Ceramic Technology, vol. 5, no. 3, pp. 234–242, 2008.

37. Y. S. Liao, H. C. Li, and Y. Y. Chiu, "Study of laminated object manufacturing with separately applied heating and pressing," International Journal of Advanced Manufacturing Technology, vol. 27, no. 7-8, pp. 703–707, 2006.

38. B. Vaupotic, M. Brezocnik, and J. Balic, "Use of PolyJet technology in manufacture of new product,"Journal of Achievements in Materials and Manufacturing Engineering, vol. 18, no. 1-2, pp. 319–322, 2006.

39. R. Singh, "Process capability study of polyjet printing for plastic components," Journal of Mechanical Science and Technology, vol. 25, no. 4, pp. 1011–1015, 2011.

40. V. Petrovic, J. Vicente, H. Gonzalez et al., "Additive layered manufacturing: sectors of industrial application shown through case studies," International Journal of Production Research, vol. 49, no. 4, pp. 1061–1079, 2011.

41. K. U. Bletzinger and E. Ramm, "Structural optimization and form finding of light weight structures,"Computers and Structures, vol. 79, no. 22–25, pp. 2053–2062, 2001.

42. A. Williams, "Architectural modelling as a form of research," Architectural Research Quarterly, vol. 6, no. 4, pp. 337–347, 2002.

43. SweetOnionsCreations,"Architecturemodeland3Dprinting—sweetonion creations," 2007,http://www.youtube.com/watch?v=rEzugxybKmA.

44. M. Phair, "Rapid prototyping: the next wave in architectural modeling," Building Design & Construction, vol. 45, no. 11, pp. 15–16, 2004.

45. I. Gibson, T. Kvan, and W. Ling, "Rapid prototyping for architectural models," Rapid Prototyping Journal, vol. 8, no. 2, pp. 91–99, 2002.

46. J. Giannatsis, V. Dedoussis, and D. Karalekas, "Architectural scale modelling using stereolithography,"Rapid Prototyping Journal, vol. 8, no. 3, pp. 200–207, 2002.

47. F. Rengier, A. Mehndiratta, H. von Tengg-Kobligk et al., "3D printing based on imaging data: review of medical applications," International Journal of Computer Assisted Radiology and Surgery, vol. 5, no. 4, pp. 335–341, 2010.

48. W. J. James, M. A. Slabbekoorn, W. A. Edgin, and C. K. Hardin, "Correction of congenital malar hypoplasia using stereolithography for presurgical planning," Journal of Oral and Maxillofacial Surgery, vol. 56, no. 4, pp. 512–517, 1998.

49. Chaput, Christophe, and J. B. Lafon, "Ceramic industry," vol. 161, no. 9, pp. 15–16, 2011.

50. G. Fielding, A. Bandyopadhyay, and B. Susmita, "Effects of silica and zinc oxide doping on mechanical and biological properties of 3D printed tricalcium phosphate tissue engineering scaffolds," Dental Materials, vol. 28, no. 2, pp. 113–122, 2012.

51. J. Suwanprateeb, R. Sanngam, W. Suvannapruk, and T. Panyathanmaporn, "Mechanical and in vitro performance of apatite-wollastonite glass ceramic reinforced hydroxyapatite composite fabricated by 3D-printing," Journal of Materials Science, vol. 20, no. 6, pp. 1281–1289, 2009.

52. R. Makovec, "Digital technologies in dental laboratories," Annals of DAAAM & Proceedings, p. p1579, 2010.

53. R. van Noort, "The future of dental devices is digital," Dental Materials, vol. 28, no. 1, pp. 3–12, 2012.

54. A. Moscaritolo, "Woman receives 3D printer? created transplant Jaw," PC Magazine Online, 2012,http://www.pcmag.com/article2/0,2817,2399887,00.asp.

55. S. J. Hollister, "Porous scaffold design for tissue engineering," Nature Materials, vol. 4, no. 7, pp. 518–524, 2005.

56. M. A. Stoodley, J. R. Abbott, and D. A. Simpson, "Titanium cranioplasty using 3-D computer modelling of skull defects," Journal of Clinical Neuroscience, vol. 3, no. 2, pp. 149–155, 1996.

57. N. Herbert, D. Simpson, W. D. Spence, and W. Ion, "A preliminary investigation into the development of 3-D printing of prosthetic sockets," Journal of Rehabilitation Research and Development, vol. 42, no. 2, pp. 141–146, 2005.

58. B. Christensen, "New device prints human tissue," 2009, http://www.livescience.com/5977-device-prints-human-tissue.html.

59. T. Qian and Y. Wang, "Micro/nano-fabrication technologies for cell biology," Medical and Biological Engineering and Computing, vol. 48, no. 10, pp. 1023–1032, 2010.

60. S. J. Song, J. Choi, Y. D. Park et al., "Sodium alginate hydrogel-based bioprinting using a novel multinozzle bioprinting system," Artificial Organs, vol. 35, no. 11, pp. 1132–1136, 2011.

61. V. Mironov, N. Reis, and B. Derby, "Bioprinting: a beginning," Tissue Engineering, vol. 12, no. 4, pp. 631–634, 2006.

62. M. Conner, "3-D medical printer to print body parts," EDN, vol. 55, no. 3, p. 9, 2010.

63. L. Zhao, V. Lee, S. Yoo, G. Dai, and X. Intes, "The integration of 3-D cell printing and mesoscopic fluorescence molecular tomography of vascular constructs within thick hydrogel scaffolds,"Biomaterials, vol. 33, no. 21, pp. 5325–5332, 2012.

64. "A step forward for artificial blood vessels," Expert Review of Cardiovascular Therapy 5.5, 817+, Academic OneFile, 2007.

65. J. Thilmany, "Printed life: the 3-D printing of living organs for transplant isn›t far-fetched, it›s almost here," Mechanical Engineering-CIME, 44+, Academic OneFile, 2012.

66. M. R. Dusseiller, D. Schlaepfer, M. Koch, R. Kroschewski, and M. Textor, "An inverted microcontact printing method on topographically structured polystyrene chips for arrayed micro-3-D culturing of single cells," Biomaterials, vol. 26, no. 29, pp. 5917–5925, 2005.

67. R. D. Piner, J. Zhu, F. Xu, S. Hong, and C. A. Mirkin, "‹Dip-pen› nanolithography," Science, vol. 283, no. 5402, pp. 661–663, 1999.

68. T. Wohlers, "Making products by using additive manufacturing," Manufacturing Engineering, vol. 146, no. 4, pp. 70–74, 2011.

69. J. Malik, "Are 3D-printed fabrics the future of sustainable textiles?" in Ecouterre, 2010,http://www.ecouterre.com/are?3d?printed?fabrics?the?future?of?sustainabletextiles/.

70. F. Tortul, "3D printed shoes awarded most creative at Mittelmoda," in Freedom of Creation, 2010,http://www.freedomofcreation.com/for/3d-printed-shoes-awarded-most-creative-at-mittelmoda.

71. P. Wallich, "3-D Printers Proliferate [Hands On]," IEEE Spectrum, vol. 47, no. 9, p. 23, 2010.

72. T. Bradshaw, "The printer that is transforming the world of creation," The Bath Chronicle, p. 45, 2011.

73. M. Simkins, "3-D printing a goo goo: conceive and build your own tasty designs with sticky food or clay," Technology and Leanrning, p. 48, 2008.

74. Singh and Sandeep, Beginning Google SketchUp for 3D Printing, Springer, 2010.

Chapter 5

A NOVEL MANUFACTURING TECHNOLOGY FOR RF MEMS DEVICES ON CERAMIC SUBSTRATES

V. Schirosi, G. Del Re, L. Ferrari, P. Caliandro, L. Rizzi, and G. Melone

Microelectronic Research, OPTEL InP Consortium Microelectronic Research Lab, c/o Cittadella della Ricerca, S.S. 7 Km 7.3, 72100 Brindisi, Italy

ABSTRACT

Microelectromechanical systems are often used for their enormous capability and good qualities in T/R modules especially for space modular applications. High isolation and very low insertion loss are guaranteed by their intrinsic working principle. This is a very robust, flexible, and low-cost technology, and it provides high reliability, good reproducibility, and complete fulfillment of technical requirements.

INTRODUCTION

The exploitation of MEMS technology for RF applications enables the manufacturing of high-performance and low-cost lumped components like variable capacitors, inductors, and switches. Starting from these elements, the implementation in MEMS technology of complete subnetworks to be integrated within RF and microwave functional blocks (e.g., RF transceivers) leads to a wider reconfigurability and, consequently, operability of the whole system.

So, RF MEMS technology provides advanced solutions to fabricate very interesting devices for switching of radio frequency signals. This technology allows for reducing drastically device dimensions and number of connections, to operate with signal in large bandwidth, from 0 to 50 GHz, with consistent performances. MEMS devices, in fact, have been identified as a promising enabling technology thanks to their extremely reduced dimensions, their high isolation in open-circuit state, and low insertion loss in short-circuit state, the high linearity. Also, monolithic integration of MEMS devices with transmission lines on substrates with high dielectric constant such as ceramic substrates (aluminum oxide, aluminum nitride, LTCC, etc.) allows manufacturing of complex devices (phase shifters, power dividers, tunable filters, couplers, reconfigurable antennas, etc.) for phase and amplitude modulation which are essential components of TX/RX modules [1, 2].

RF MEMS DESIGN

The RF MEMS devices described have been full designed, fabricated, and tested at Microelectronic Research Lab of Optel Consortium.

Design is carried out by electromechanical and electromagnetic simulation of RF MEMS. Electromechanical design consists of an FEM modeling of micromechanic structures. The goal of the design is the optimization of parameters such as switching time, overshoot, and pull-in voltage. FEM simulations provide an actuation time of 10–15 μsec, an overshoot less than 13% of the gap and a pull-in voltage of 15 V [3–5].

Electromagnetic design consists of full-wave simulations that provide S parameter curves of the device [6, 7]. For a shunt switch, in OFF state, return loss is better than −25 dB and insertion loss is about −0.1 dB, in the range 0–20 GHz. In ON state, return loss is about −0.1 dB and isolation is better than −30 dB in the range 9.5–16 GHz and, in particular, isolation is higher than −45 dB at resonance frequency (14 GHz).

RF MEMS devices are designed in Coplanar Waveguide or Microstrip technology, and every device geometry can be adapted to any kind of operating bandwidth in the 1 : 50 GHz range.

FABRICATION ON CERAMIC SUBSTRATE

RF MEMS fabrication requires only a surface micromachining technology. Optel technology is independent of the substrate used since it requires just thin-film PVD metal deposition, CVD passivation, and gold electroplating.

Optel technology enables the manufacturing of reliable RF MEMS devices onto different substrates:(i)Si (for consumer electronic and integration with CMOS technology),(ii)GaAs (for III–V semiconductor technology and subsequent integration in high frequency applications),(iii)GaN/Si (for monolithic integration with active high-power/high-speed amplifying electronics such as GaN-HEMT).

Also, in addition to semiconductor substrates, RF MEMS devices have been monolithically manufactured on ceramic such as alumina and LTCC polished substrates. Main advantages resulting from the use of ceramic substrates are a good thermal stability, excellent hardness and wear resistance, a good corrosion resistance, excellent dielectric property, and acceptable thermal conductivity. These properties make this substrate suitable for space applications.

RF MEMS devices have been monolithically manufactured on 300 μm thick alumina (2 inches). The entire process consists of about 100 steps, and it requires 8 photolithographic levels. It is a very robust, flexible, and low-cost technology.

Robustness of the manufacturing process is due to many improvements introduced by Optel technology, both at electrostatic and electromagnetic levels, with respect to MEMS devices on semiconductors. First of all, a ceramic substrate, because, it's intrinsic high isolation, unlike a semiconductor, does not take part in MEMS operation. Also, a ceramic substrate is not influenced by ionizing radiations, so it avoids many problems of charge generations in space applications. Finally, charge trapping phenomenon, one of the most important causes of failure for RF MEMS on semiconductors, can be completely avoided, just because dielectric is not needed on high-isolation substrate. In fact, removing the dielectric underneath the beam, as well as introducing dimples anchored on the substrate or integrated into the beam to stop the bridge actuation, allows the elimination of the insurgence of charging trapping phenomena and, in addition, allows even the reduction of the pull-in voltage needed for membrane deflection.

Also, Optel technology is low-cost technology, with respect to well-established RF MEMS manufacturing on silicon. In fact, a semiconductor substrate is more expensive than a ceramic substrate, and the technology proposed is a process for microelectronic passive devices. This is allowed to done without the process of active IC such as doping, ionic implantation, and oxide growth.

In the standard proposed process for RF MEMS devices, there are 4 metallization layers: a resistive, a conductive, and 2 galvanic thickness layers. They are separated by 2 passivation layers where through vias are opened to provide interconnections with underneath layers. The sacrificial layer is a photoresist upon which the membranes are defined by a galvanic growth.

As first step, a 600-nm thick silicon nitride is deposited as an insulating layer and a high-resistance metal is deposited and defined to create actuation pads and bias lines. Next, a new silicon nitride layer is deposited to provide the high isolation needed for the actuation electrodes. Contact DC vias are then defined and etched within the silicon nitride layer. A multimetal underpass is deposited and defined to create RF lines underneath the bridge, and it is covered by a second passivation layer, which provides an insulating layer for RF lines. Then, RF vias are opened within silicon nitride. Next, a gold layer is deposited and defined in order to provide low-resistance electrical contacts. The sacrificial layer needed for definition of a suspended beam is composed of a 3 μm thick photoresist. An Au-based multimetal layer is deposited to obtain the electrical continuity layer, previously to galvanic gold electroplating. Then, a 1 μm thick gold layer is grown to define the membranes and a 2 μm thick gold layer is grown to define RF lines. Finally, sacrificial layer is removed by a plasma etching process (Figures 4 and 5).

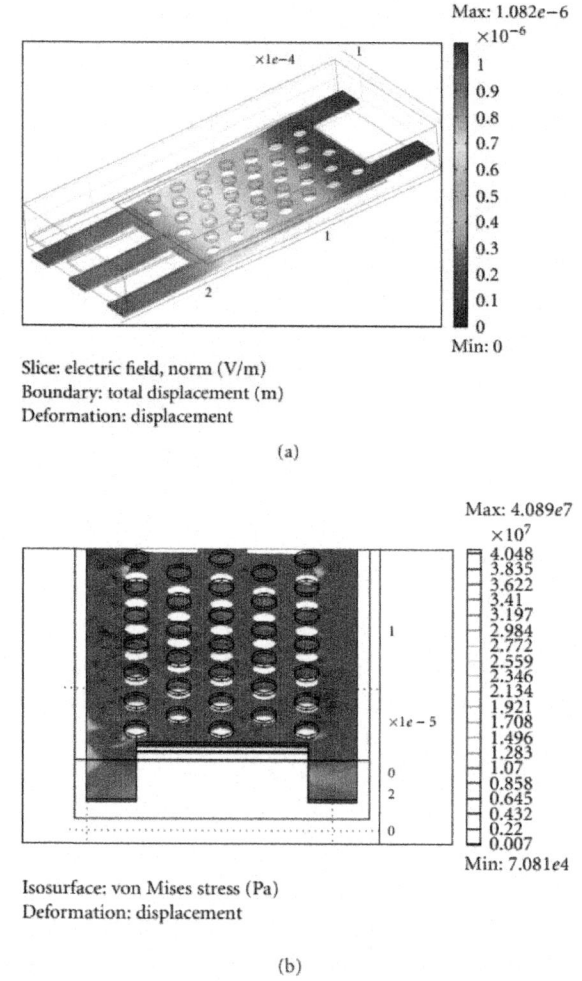

Slice: electric field, norm (V/m)
Boundary: total displacement (m)
Deformation: displacement

(a)

Isosurface: von Mises stress (Pa)
Deformation: displacement

(b)

Figure 1: Simulated actuation for a shunt switch membrane (a) and simulated stress concentration on the anchor zone (b).

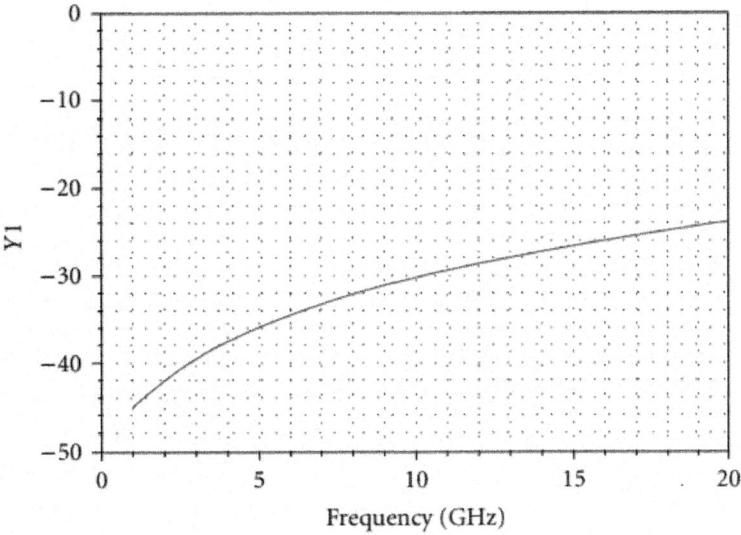

Figure 2: Simulated return loss (blue curve) and insertion loss (red curve) for a shunt switch in OFF state.

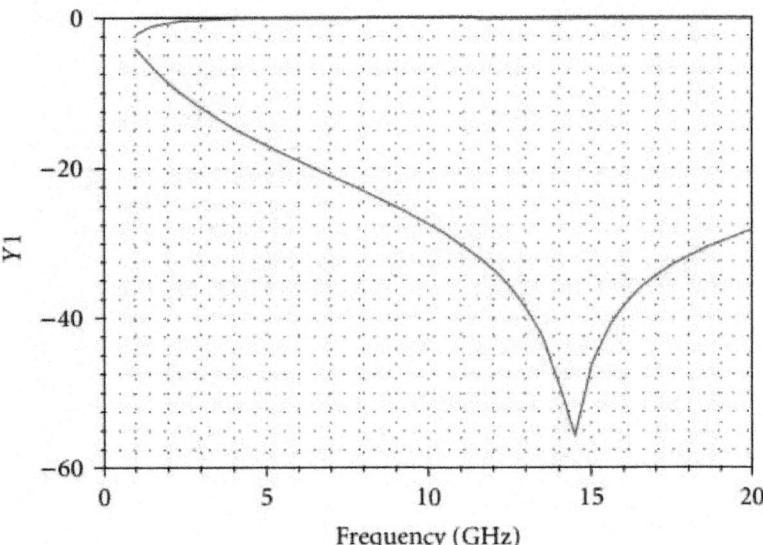

Figure 3: Simulated return loss (blue curve) and isolation (red curve) for a shunt switch in ON state.

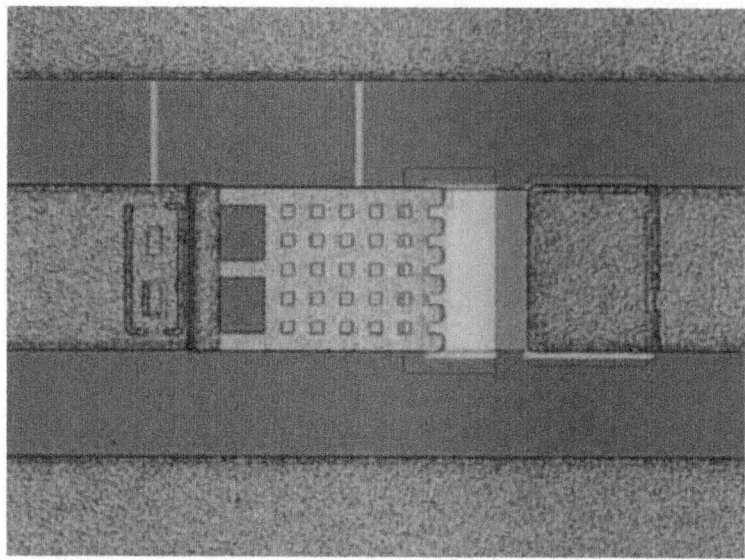

Figure 4: Cantilever RF MEMS switch manufactured on alumina substrate.

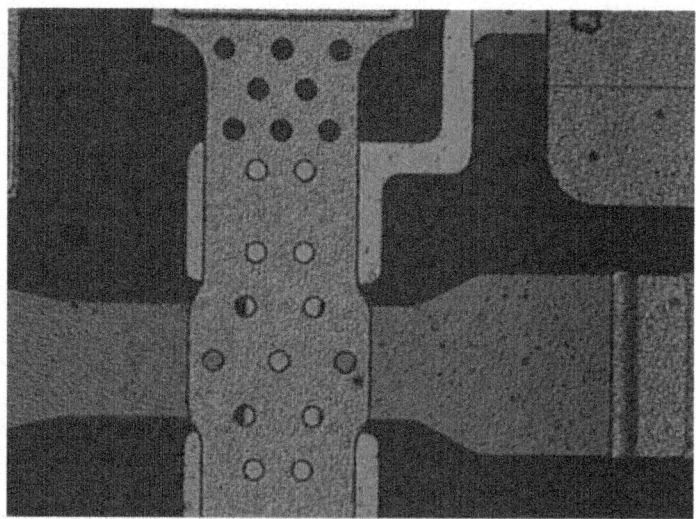

Figure 5: Fixed-fixed serie RF MEMS switch manufactured on LTCC substrate.

For a manufacturing process on ceramic substrate, the passivation layers can be reduced only to small areas where an isolation between DC and RF lines is needed and can be completely removed underneath the bridge.

Critical steps are both membrane definition and sacrificial layer removing. In particular, the reduction of ion bombardment on membranes is a well-known problem that could lead to a stressed and deformed membrane after the dry etching process. In fact, sacrificial layer etching requires a highly confined plasma to minimize the damage on metallic suspended structures. So, introducing appropriate changes in process conditions, the result has been good. The beams are characterized by very low stress, high flatness, and absence of damage caused by reduced thermal and ion collision during the plasma etching process. Furthermore, mechanic behavior is unchanged both for semiconductors and for bulk or multilayer ceramics (Figure 6).

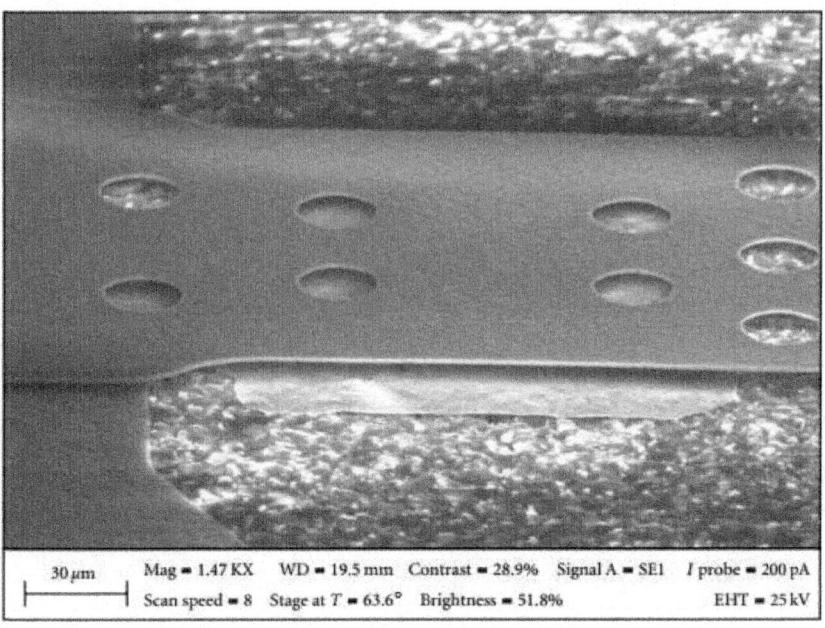

Figure 6: SEM view of a suspended beam on LTCC substrate. Beam is characterized by very low stress, high flatness, and no damage.

MEASUREMENTS

Testing is carried out on single RF Switch and on RF MEMS microwave circuits. Testing steps include the following:(i)static DC testing (for measurement of pull-in voltage, contact resistance, and capacitance ratio)(ii)dynamic DC testing (for measurement of switching times and lifetime testing)(iii)RF testing (for measurement of return loss, insertion loss, isolation, and phase shift).

Static DC testing is implemented with voltage application through the bias DC pads. Capacitance ratio is measured for shunt switches, and it is obtained using a C-V meter, which allows the capacitance measurement related to the RF line under the bridge. Optel switches have an up-state capacitance of 0.170 pF and a down-state capacitance of 5.4 pF. The capacitance ratio is 31.76. Contact resistance, for a serie switch, is measured using a multimeter between the suspended bridge and the underpass. Ohmic switches on ceramic substrates have a contact resistance of 6 Ω. This high value is due to the ceramic substrate roughness that prevents a well-defined contact area between the suspended membrane and the RF line. A method to obtain a well-defined contact area is under investigation.

Dynamic measures are implemented using a custom setup consisting of a signal generator which drives a high slew rate operational amplifier with high voltage supply, for generation of pulse actuation train (Figure 7).

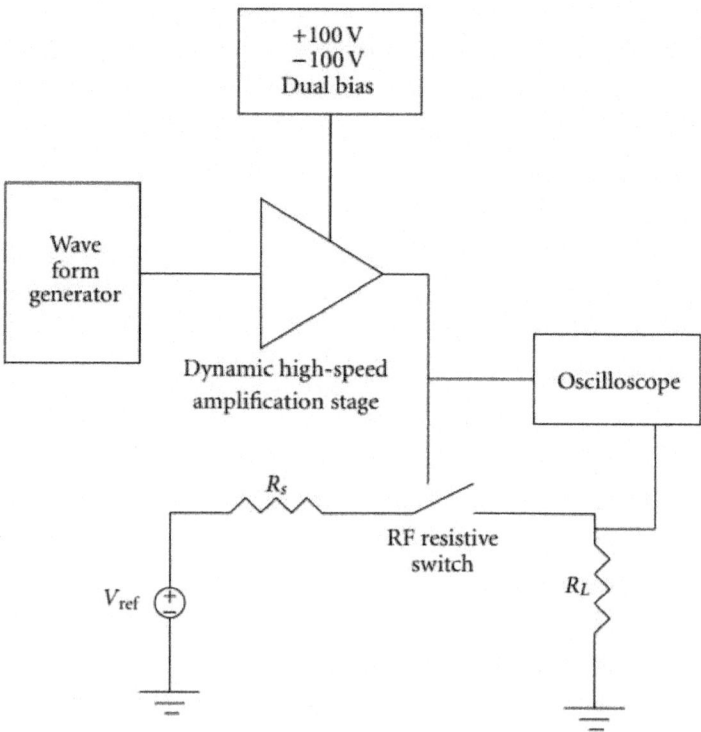

Figure 7: Custom setup of dynamic behaviour measurements for cantilever ohmic switches.

The dynamic measurements are possible with two channels oscilloscope at 10 GS/s of sample rate, through the serial circuit (Figure 8), with 1 V voltage supply across the open circuit line [10]. Serie switches show pull-in voltage of 12 V and actuation and deactuation time of 6 and 14 μsec. Cantilever topology shows better results than fixed-fixed one.

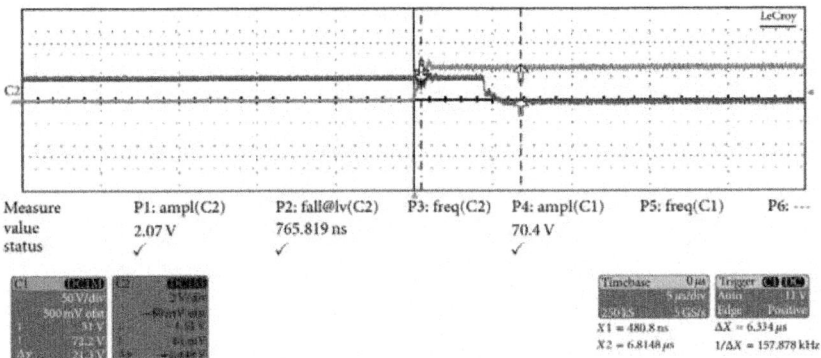

Figure 8: Oscilloscope screenshot: measurement of actuation time for cantilever ohmic switch.

Electrical endurance lifetime measures are carried out by the same way of dynamic testing, applying a pulse train to obtain 10^8 actuations of the switches. After 100 million actuation cycles, switches show an actuation and deactuation time drift (30–50 μsec) and a rising pull-in voltage (70 V). Anyway, no structural fail has been detected on switches after 10^8 cycles.

RF testing is implemented using a parametric network Analyzer for measurements of S parameters. In the following graphics, measured return loss, insertion loss and isolation curves are represented. For cantilever switch, isolation is always better than −20 dB and insertion loss is about −0.4 dB on the frequency band 1–50 GHz (Figure 9). For shunt switches, measured insertion loss is always better than −0.25 dB and isolation is about −33 dB at resonance frequency (Figure 10).

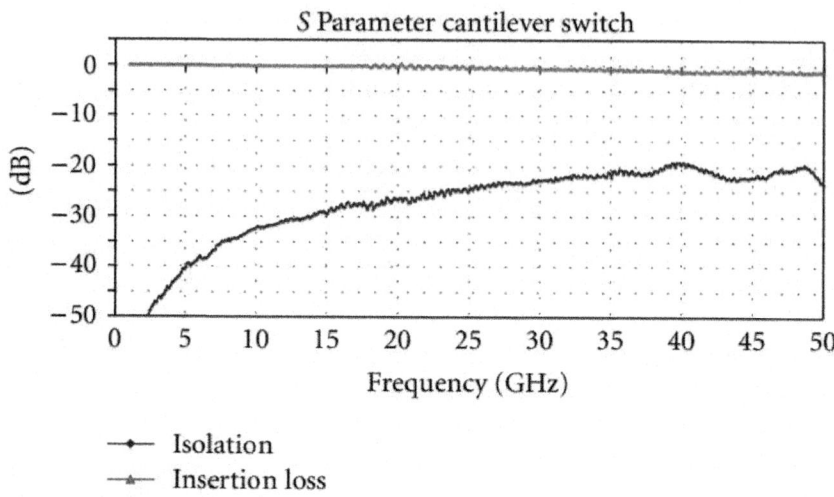

Figure 9: Measured insertion loss and isolation of a cantilever switch on alumina substrate.

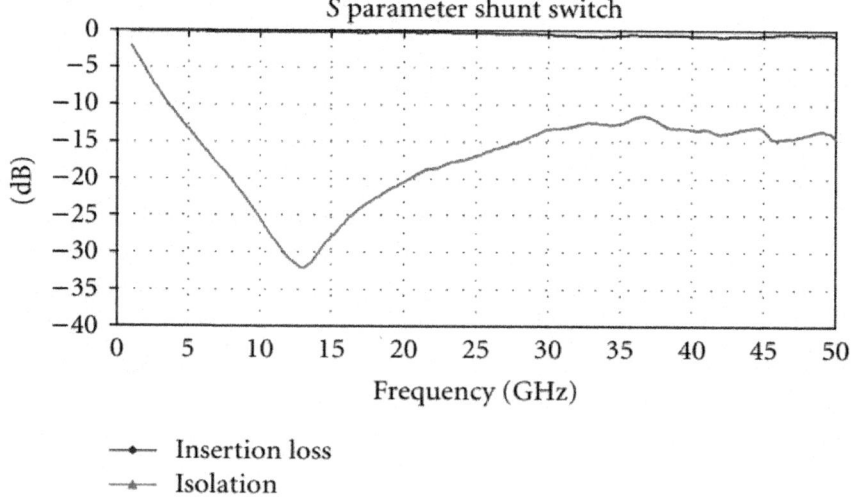

Figure 10: Measured insertion loss and isolation of a Shunt switch on alumina substrate.

Finally, in Table 1, a comparison of performances for three kinds of RF MEMS switches is shown, but presented performances are not closely comparable because they refer to developing research prototypes. Anyway, about Optel switches, cantilever topology has shown the best trade-off between DC and RF performances. Typically, the main disadvantage of ohmic contact MEMS switches is that they show higher loss with respect to capacitive switches due to the nonzero contact resistance between the bridge and the transmission line, and contact resistance for ceramic substrates is further increased because of an intrinsic roughness of the ceramic. In addition, the contact area is liable to high current density and possible material transfer which can lead to untimely failure, but, on other hand, cantilever switch allows a reduced actuation time and a reduced pull-in voltage with respect to double-anchored switch (fixed-fixed serie and shunt). These aspects have been accurately analyzed and optimized in order to obtain a compact low-loss ohmic contact cantilever MEMS switch.

Table 1: Performance measured on RF MEMS switches manufactured in Optel, Fondazione Bruno Kessler, and RADANT MEMS.

	Optel			University of Perugia-FBK [8]	RADANT MEMS [9]
	Cantilever	Serie fixed-fixed	Shunt	Cantilever (*)	Cantilever (*)
Pull-in voltage [V]	12	30	45	38	40–120
Contact resistance [Ω]	6	12	//	1.48	1
$C_{ratio} = C_{down}/C_{up}$	//	//	32	//	//
Switching time [us]	6	14	18	n.a.	5
Handled power [dBm]	−10	−10	−10	30	27
Return loss [dB]	>25 dB up to 20 GHz	>18 dB up to 20 GHz	>20 dB up to 20 GHz	>30 dB up to 20 GHz	>20 dB up to 36 GHz
Insertion loss [dB]	<0.8 dB up to 20 GHz	<0.8 dB up to 20 GHz	<0.25 dB up to 20 GHz	<0.3 dB up to 20 GHz	<0.5 up to 38 GHz
Isolation [dB]	>25 dB up to 20 GHz	>25 dB up to 20 GHz	>20 dB at 8–20 GHz, >33 dB at resonance	>20 dB up to 13 GHz >10 dB up to 40 GHz	20 dB @ 10 GHz, 13 dB @ 40 GHz
Lifetime [*cycles > 10^8]	no failure	no failure	no failure	no failure	no failure

(*) switches manufactured on High Resistivity Silicon.

So, the designed cantilever switch has become a building block of reconfigurable RF MEMS devices. In fact, Optel activity is directed to design devices used in T/R modules for high-frequency communications, such as step delay modules, phase shifters, and power dividers. This kind of devices requires low insertion loss, high reconfigurability, and small size. RF MEMS devices represent an extremely attractive alternative to provide this requirements this is possible because RF MEMS devices guarantee low loss, low-power consumption, and excellent linearity compared to the traditional MMIC. Also, high-frequency communications require both transmitting and receiving antenna systems with an electronic beam steering, and RF MEMS can be used in a phase shifter to control the phase of the individual radiating element of the antenna array.

CONCLUSIONS

In conclusion, switches have shown a good matching between simulations and measurements. The manufacturing technology is very robust, flexible, and low cost, and it provides high reliability, good reproducibility, and complete fulfillment of technical requirements.

Currently, our team is working on future improvements for RF MEMS integration with signal conditioning circuits on ceramics substrates; other developments are RF packaging and interfacing with commercial connectors, a design optimization of switches and complex devices in particular for power handling improvement, ageing, and life time tests in harsh environmental conditions for MIL-std fulfillment, qualification for space applications and satellite communications.

ACKNOWLEDGMENTS

This research activity is partially funded by National Research Program of MIUR no. 800/04 entitled "Enabling Technologies for Microwave Tx/Rx Systems".

REFERENCES

1.   J. Bouchaud and H. Wicht, "RF MEMS: status of the industry and roadmaps," in Proceedings of the IEEE Radio Frequency Integrated Circuits (RFIC) Symposium—Digest of Papers, pp. 379–384, June 2005.

2.   G. M. Rebeiz, RF MEMS, Theory, Design and Technology, John Wiely & Sons, New York, NY, USA, 2002.

3.   S. Chen, T. V. Baughn, Z. J. Yao, and C. L. Goldsmith, "A new in situ residual stress measurement method for a MEMS thin fixed-fixed beam structure," Journal of Microelectromechanical Systems, vol. 11, no. 4, pp. 309–316, 2002.

4.   Y. Pauleau, "Generation and evolution of residual stresses in physical vapour-deposited thin films,"Vacuum, vol. 61, no. 2–4, pp. 175–181, 2001.

5.   Z. Wang, L. Chow, J. L. Volakis, K. Saitou, and K. Kurabuyashi, "Contact physics modeling and optimization design of RF-MEMS cantilever switches," in Proceedings of IEEE Antennas and Propagation Society International Symposium and USNC/URSI Meeting, vol. 1A, pp. 81–84, July 2005.

6.   S. C. Saha, T. Singh, and T. Sæther, "Design and simulation of RF MEMS cantilever and bridge switches for high switching speed and low voltage

operation and their comparison," in Proceedings of the International Symposium on Signals, Circuits and Systems (ISSCS '05), pp. 131–134, July 2005.

7.   X. Rottenberg, B. Nauwelaers, R. Mertens, et al., "RFMEMS metal contact capacitive switches," inProceedings of the 4th Round Table on Micro/Nano Technologies for Space, Estec, Noordwijk, The Netherlands, May 2003.

8.   G. de Angelis, A. Lucibello, E. Proietti, et al., "RF MEMS ohmic switches for matrix configurations," inProceedings of the 11th International Symposium on RF MEMS and RF Microsystems (MEMSWAVE '10), Otranto, Italy, June 2010.

9.   RADANT MEMS, http://www.radantmems.com/.

10. C. Calaza, B. Margesin, F. Giacomozzi, K. Rangra, and V. Mulloni, "Electromechanical characterization of low actuation voltage RF MEMS capacitive switches based on DC CV measurements," Microelectronic Engineering, vol. 84, no. 5–8, pp. 1358–1362, 2007.

Chapter 6

COMPLEX METALLIC ALLOYS AS NEW MATERIALS FOR ADDITIVE MANUFACTURING

Samuel Kenzari, David Bonina, Jean Marie Dubois, and Vincent Fournée

Institut Jean Lamour, UMR 7198 CNRS-Université de Lorraine, F-54011 Nancy, France

ABSTRACT

Additive manufacturing processes allow freeform fabrication of the physical representation of a three-dimensional computer-aided design (CAD) data model. This area has been expanding rapidly over the last 20 years. It includes several techniques such as selective laser sintering and stereolithography. The range of materials used today is quite restricted while there is a real demand for manufacturing lighter functional parts or parts with improved functional properties. In this article, we summarize recent work performed in this field, introducing new composite materials containing complex metallic alloys. These are mainly Al-based quasicrystalline alloys whose properties differ from those of conventional alloys. The use of these materials allows us to produce light-weight parts consisting of either metal–matrix composites or of polymer–matrix composites with improved properties. Functional parts using these alloys are now commercialized.

INTRODUCTION

Additive manufacturing technologies were first used for rapid prototyping but are now increasingly being used to produce a series of end user functional parts. It consists in producing a three-dimensional (3D) object, usually layer by layer, from a single computer file (a CAD model). These technologies are opposed to more traditional subtractive methods, such as machining, which proceed by removal of material to obtain the 3D object. These new manufacturing methods are becoming widespread and affect a lot of industries (automotive, aerospace, machinery, medical and dental, design, etc). They generated global revenues of \$2.204 billion in 2012 and the average annual growth over the 2010–2012 period was around 28% [1].

The selective laser sintering (SLS) is one such method developed and patented by Deckard *et al* [2]. The process used in this study operates as follows. The object is first modeled in a CAD file and decomposed into as many two-dimensional (2D) layers as necessary, with a thickness of the order of 100 μm. The layers are built successively according to the principle illustrated in figure 1. A first layer of powder is spread by an automated roll on a platform and an infrared laser is used to convert the powder to a solid object by ‹selective sintering› without external pressure according to the 2D plot defined by the CAD model. The powder used contains at least one polymer whose melting temperature is generally about 200 °C. The powder bed is preheated a few degrees below its melting point and the laser just provides locally the thermal energy required to bring the polymer to its melting temperature. Un-melted powders naturally provide support for the following layers. Then the build platform is moved down by the thickness of one layer, a new layer of powder is spread onto the previous one and the cycle repeats itself to build the part from bottom to top.

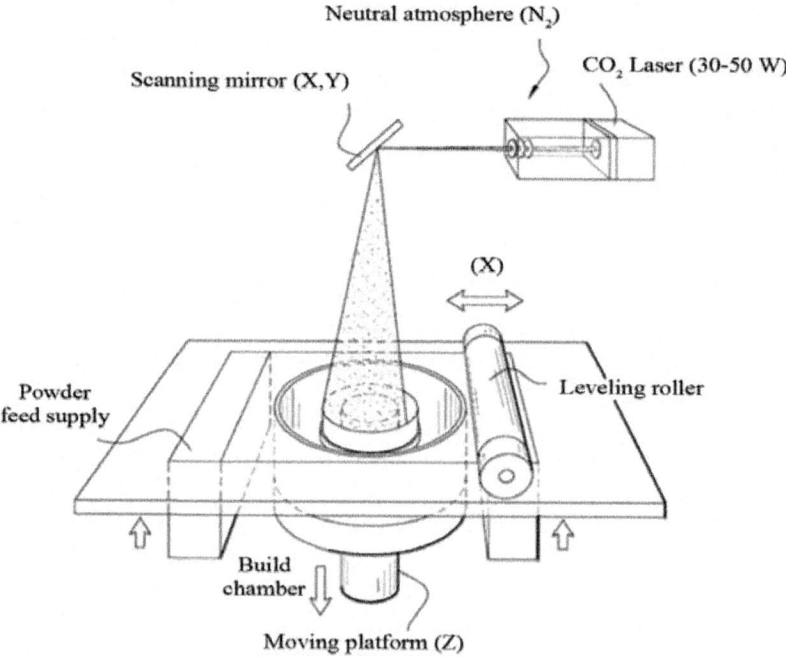

Figure 1: Schematic description of the SLS process used. The laser builds 2D section of the parts according to the CAD model. Then the platform is lowered and another

layer of powder is spread onto the previous one. The process is repeated until the 3D part is completed. Un-melted powders serve as support for the next layers.

The SLS process is used to construct parts that are either entirely made of polymer or a polymer composite reinforced by different kinds of particles or even a metal matrix composite [3, 4]. The latter is referred to as indirect SLS because it requires a two-step process. First, a porous preform must be obtained by SLS of a powder mixture containing metal particles and a polymer binder. At this stage, the object is a metal/polymer composite containing 40–50% of pores by volume. It is rigid enough to be handled but has no sufficient mechanical strength to be used directly. In a second step, this preform is submitted to a heat treatment during which the polymer is debinded and simultaneously infiltrated by a filler metal whose melting point is necessarily lower than that of the metal used in the preform. Infiltration of the porous preform with the liquid filler metal occurs by capillarity. This two-step process leads to composite metal parts made entirely of the metal base and the filler metal. In practice, the indirect SLS is a well-controlled process only for steel/brass composites.

Since the year 2000s, a series of studies was conducted to extend the materials range for indirect SLS, and especially towards light alloys like aluminum alloys [5]. Today, these developments are still at the stage of being a technological challenge. Aluminum is also used in direct SLS, incorporated into a polymer matrix to obtain composite parts with a metallic appearance [6]. Basically, SLS polymer composites have a high level of porosity which prohibits their direct use in applications requiring a perfect seal at low thicknesses of parts without post-impregnation of resin. This sealing problem exists also for all other reinforcing materials used in direct SLS, not just aluminum. To circumvent this problem, reinforced polymer composite parts must undergo a second processing step of impregnation with a liquid resin to fill the porosity. In addition, the friction and wear properties of composite parts produced by direct SLS are relatively poor, which limits their use for movable parts.

The work presented in this article shows how some of the problems mentioned above could be solved through the introduction of complex metallic alloys (CMAs) as new materials for rapid manufacturing technologies. In the case of indirect SLS, it is possible to obtain fully functional metal parts made of different light alloys (Al-based). In the case of direct SLS, CMA particles are used as reinforcement particles in a polymer matrix to improve the properties of composite parts. CMA particles can also be used as reinforcement particles in resin composites made by stereolithography (SL).

CMAs

CMAs are defined as intermetallic compounds with large unit cells, containing several tens up to several thousands of atoms. Inside the unit cell, atoms usually form geometrical cluster units of high symmetry, for example icosahedral (five-fold symmetry) or decagonal (ten-fold symmetry) clusters. These clusters are interconnected by so-called glue atoms to develop a 3D structure. There has been a renewed interest in such complex compounds since the discovery of quasicrystals (QCs) by Dan Shechtman in 1982 [7]. QCs can be viewed as the ultimate degree of complexity in CMAs, with a unit cell of infinite dimension and the loss of the translational invariance characterizing conventional crystals. QCs first attracted attention due to their unique atomic structure, inducing a paradigm shift in crystallography. In parallel, intriguing physical properties were reported, not expected for intermetallic compounds (for a recent review, see [8]). Among them, the thermal and electronic transport properties were found to be non-metallic in several systems like Al–Cu–Fe QCs, although they are made of metals.

From the point of view of technological applications, CMAs have often been presented as promising new materials because of their potentially useful properties like low coefficients of friction, relatively good corrosion resistance, low wettability, good wear resistance, etc [9]. Such functional properties are however difficult to implement due to the intrinsic brittleness of CMAs which prevents their use as bulk materials. Two main alternatives have been considered to circumvent this problem. One is to use these phases as reinforcement particles in a ductile matrix like Al-based [10, 11] or a polymer matrix [12]. The other is to use CMAs as new coating materials [9, 13].

In the following sections, we will summarize and discuss the use of CMAs in metal matrix or polymer matrix composites processed by additive manufacturing technologies. The properties of resulting parts will also be described and compared to existing state of the art industrial solutions [14–18].

MANUFACTURING OF METALLIC PARTS BY INDIRECT SLS

The indirect SLS method is currently well controlled for steel/brass composites but the density of this composite (8 g cm^{-3}) is much too high for many applications. Therefore many studies have been conducted in order to extend this process to produce light-weight Al-based composites. First, a porous preform containing particles of a preliminary Al alloy of a certain grade and a polymer binder must be produced by SLS. Then the porous preform has to be infiltrated with a liquid Al alloy of a different grade characterized by a lower melting point than the Al alloy used in the preform. All attempts to produce such Al-based composites by indirect SLS have led to parts presenting poor mechanical strength [6, 19]. The origin of this brittleness is due to the nitriding of the interfaces during the debinding process, which according to [19] is a necessary step to achieve complete infiltration of the preform and maintain its dimensional accuracy.

This problem can be circumvented by manufacturing a preform based on CMA particles and a polymer binder, which can then be infiltrated with a commercial Al alloy. One key advantage of Al-based CMAs is their thermal stability (larger than 800 °C for Al–Cu–Fe–B QC for example) which is far above any commercial Al grade while still possessing low density (< 4.7 g cm^{-3}). First a suitable powder mixture is prepared by blending Al–Cu–Fe–B QC particles with polyamide (PA) (Nylon) particles. The polymer represents approximately 2.5% in weight of the total blend. The QC particles are produced by gas atomization. They have a nominal composition of $Al_{59}Cu_{25.5}Fe_{12.5}B_3$ (at.%) and contain primarily the quasicrystalline phase together with a small amount of the β-$Al_{50-x}(Cu,Fe)_{50+x}$ phase. This powder mixture is used to prepare a preform by SLS, an example being shown in figure 2. Typically, such a preform contains about 40% volume fraction of open porosity. Analysis of the microstructure by scanning electron microscopy (SEM) reveals that the PA particles are melted by the laser during SLS and form bridges between the QC particles (figure 2). The preforms are brittle at this stage but can be easily manipulated. Infiltration is realized in a pre-evacuated furnace (1.10^{-2} mbar) back-filled with an inert atmosphere. The preform is placed in a crucible on a sacrificial tab together with blocks of the infiltrating metal (figure3). In this, an Al 1050 grade was used, i.e. a commercial Al alloy having a melting point of around 660 °C. The crucible is then filled with alumina particles in order to support the preform during the debinding and infiltration process. The crucible is heated up to the melting of the infiltrant (3 h, 680 °C), which can then fill in the open porosity of the preform by capillarity.

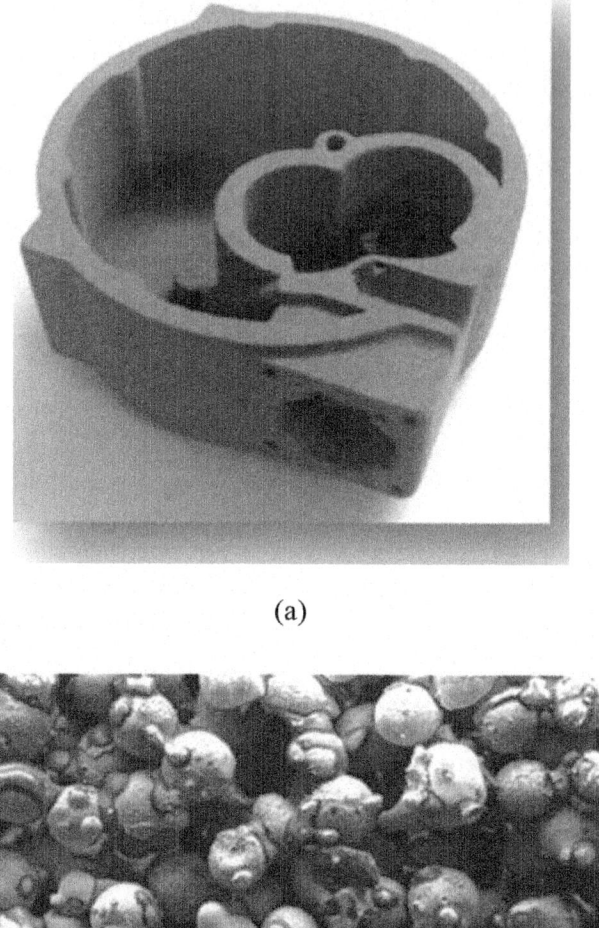

(a)

(b)

Figure 2: Left: picture of a porous preform made by SLS containing Al–Cu–Fe–B QC particles and a polymer binder. The lateral dimension of the preform is 5 cm. Right: SEM image of the mixed powder after selective sintering showing the formation of polymer bridges connecting the QC particles.

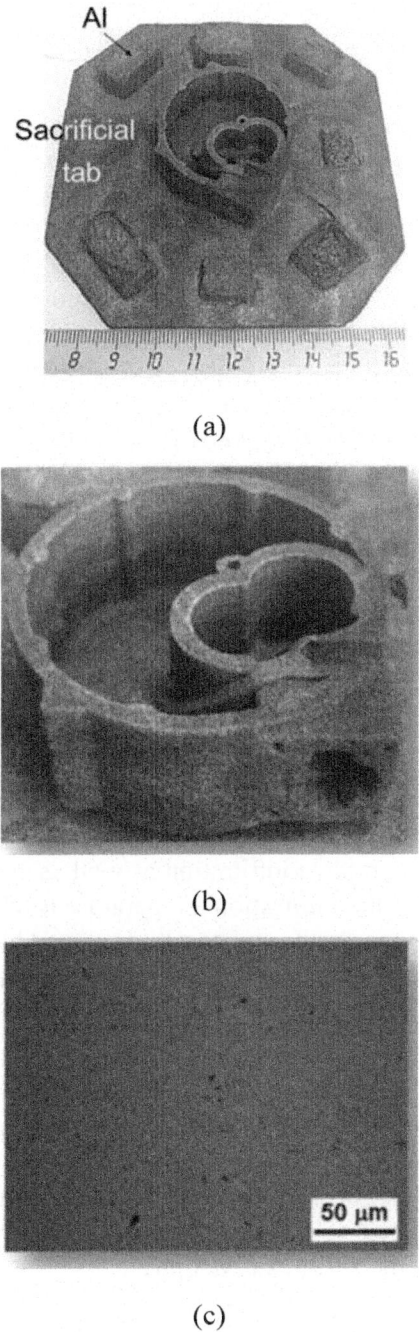

(a)

(b)

(c)

Figure 3: Left: picture of the infiltrated preform placed on the sacrificial tab surrounded by a block of Al infiltrant. Right: the top part shows a picture of infiltrated preform.

The part is now fully metallic and contains several CMA phases but no fcc Al. The bottom part shows an optical image of the infiltrated preform indicating a low level of porosity.

Figure 3 shows an example of a preform successfully infiltrated by a 1050 Al alloy. The nitriding step was not necessary in order to keep the dimensionality of the part and achieve complete infiltration. The parts contain little porosity as seen in the SEM image in figure 3, achieving values similar to those observed for steel–brass composite parts. X-ray diffraction analysis reveals that the parts are composed of several CMA phases and do not contain crystalline aluminum anymore. The absence of fcc aluminum in the final composite results from phase transformations induced by chemical diffusion of Al into the CMA particles during the infiltration cycle. The QC phase mainly transforms into tetragonal -$Al_{70}Cu_{20}Fe_{10}$, an alloy known for its beneficial contribution in Al matrix composites due to the improvement of the bonding strength between the Al matrix and the particles [20, 21]. The composite parts exhibit high hardness values (in the range of Vickers hardness 300–400 Hv), which is remarkable for Al-based materials but which makes them fragile. However it is possible to optimize the process by applying post-infiltration heat treatments and reduce the hardness down to about 200 Hv. These values are similar to those of steel/brass parts, but the final density of the QC-Al1050 composites is lower by a factor of 2, which was the targeted objective.

The bottleneck to achieve a successful technology transfer of the process relies on the ability to precisely control the thermal cycle during the infiltration in industrial conditions. It is important that the thermal treatment is applied for sufficient time to complete infiltration as well as chemical diffusion at the interfaces. However, if the infiltration temperature is too high or if the holding time is too long, then the parts may start to deform. The high melting point of Al-based QC particles is thus clearly an advantage compared to commercial Al grades because the temperature window for the infiltration of the filler metal is much larger.

To conclude this section, a new range of composite powders based on CMAs was developed to produce all-metal parts of any complex shape by indirect SLS technology [16]. The composite parts are entirely metallic; they contain a low level of porosity and have a low mass density. This method can in principle be extended to a broad spectrum of chemical compositions by varying the nature of the selected CMA phase.

MANUFACTURING OF CMA-POLYMER COMPOSITE PARTS BY DIRECT SLS

Earlier works by Bloom *et al* have shown that QC used as reinforcement particles in a polymer matrix lead to composite materials with improved wear resistance and reduced friction [22, 23]. In this section, we show that such QC-polymer composites can be adapted to SLS technology to manufacture functional 3D composite parts. The polymer matrix is made of PA 12 (Nylon) with a low density (<1.5 g cm^{-3}) and is reinforced by Al–Cu–Fe–B QC particles obtained by gas atomization representing more than 50 wt.% of the total blend [17]. The QC particles were sieved with a mesh size of 75 μm and blended with nylon particles (< 75 μm) in a Turbula mixer (30 min) in the appropriate volume fraction (30 vol.% of QC) optimized for the laser sintering process. Selective laser sintered parts were then constructed using a sPro™ 60 SLS® Center. The parameters used in SLS (laser, temperature, scanning speed, etc) have been optimized to yield dense parts, unconstrained shape, dimensional accuracy and a quality of skin similar to those of commercial composite parts normally produced by SLS. They are given in table 1. The maximum size (*xyz*) of the parts which can be produced in this experimental setup can reach 380 × 330 × 450 mm (≈ 60 l). An example is shown in figure 4. The friction, wear and sealing properties of the composite material have been studied and compared to other PA matrix composites. The main results are summarized below.

Table 1: Experimental parameters used to construct SLS parts

Laser conditions				Temperatures (°C)			Feed parameter
Power (W)	Beam diameter (mm)	Scan speed (mm s^{-1})	Scan fill spacing (mm)	Bed	Feed	Piston	Layer thickness (mm)
48	0.42	12 100	0.22	173	140	150	0.1

Figure 4: Picture of an intake manifold made of a polyamide composite reinforced by QC particles. This part is used in automotive industry and is produced by SLS in a single processing step. The size of the component is about 20 × 30 × 40 cm. Reproduced from [14], Copyright (2012), with permission from Elsevier.

Friction Properties

Friction tests were carried out at room temperature in ambient atmosphere with a relative humidity of 50–60% and under non-lubricated conditions. The indenter was a 6 mm ball of 100Cr6 hard steel. The radius of the trace was 5 mm. The sliding velocity was 15 cm s^{-1} (300 rpm) and the normal load was 10 N. The sample surfaces were first mechanically polished using SiC paper in water lubricant (from 500 grit down to 4000 grit) and cleaned with methanol and dried. The ratio between the lateral force measured using a pin-on-disk tribometer and the applied normal force gives the friction coefficient μ.

The friction curves measured on a series of composite samples prepared by SLS are shown in figure 5. The samples are PA matrix composites reinforced by either commercial fillers (carbon fibers, glass fibers, glass particles) or Al–Cu–Fe–B QC particles.

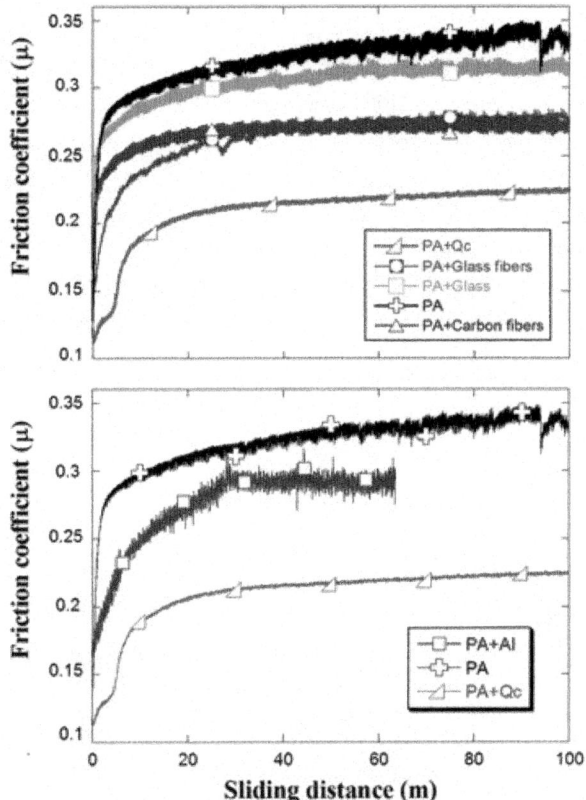

Figure 5: Top and bottom: friction coefficients measured as a function of the sliding distance for PA composites filled with various reinforcement particles or fibers. Adapted from [14], Copyright (2012), with permission from Elsevier.

The results clearly show that the lowest friction coefficients are obtained for composites reinforced by the QC particles. Other PA composites reinforced by Al particles of similar size and volume fraction than the QC particles were also tested. In this later case, all friction tests were stopped after about 65 m of sliding distance due to strong tangential forces (figure 5). SEM investigations of the worn surface (figure 6) and indenter (figure 7) show that the wear track and the ball counterpart are covered by a transferred layer which is made up primarily of aluminum and oxygen. When QCs are used as filler particles, no material transfer occurs and the surface of the composite is hardly degraded when the friction test ends.

Figure 6: SEM images of the worn surfaces of the PA composites reinforced by QC particles (left) and Al particles (right) observed after ending the friction test. Insets are optical micrographs of the sample showing the wear tracks. Adapted from [14], Copyright (2012), with permission from Elsevier.

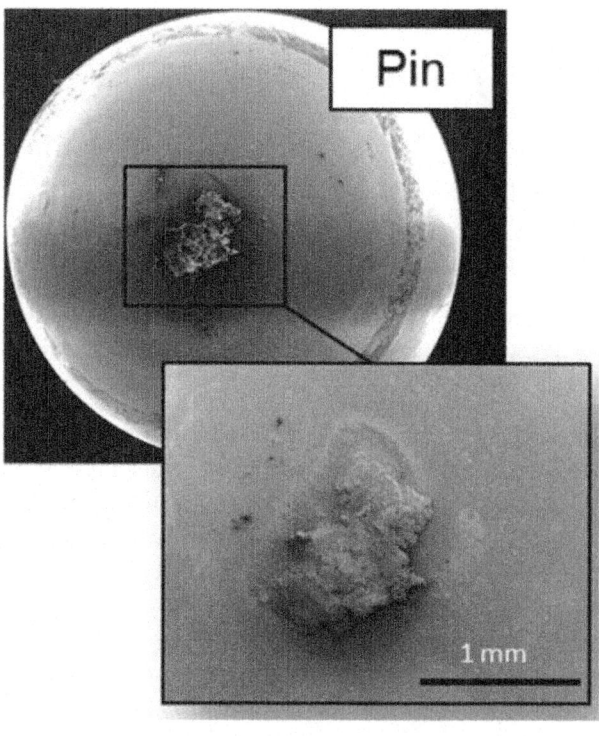

(a)

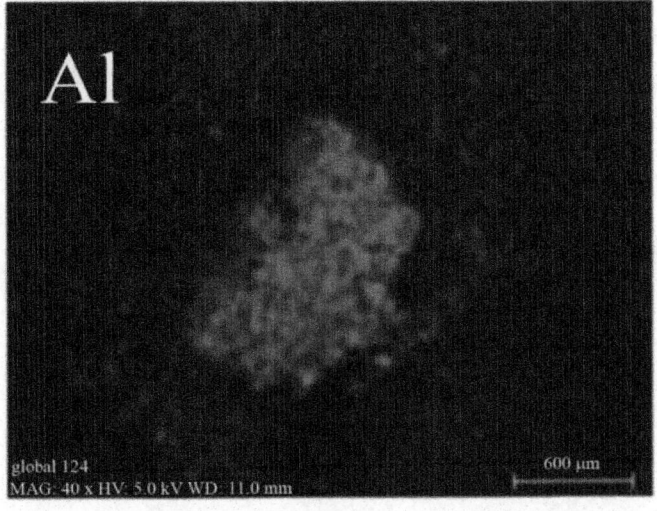

(b)

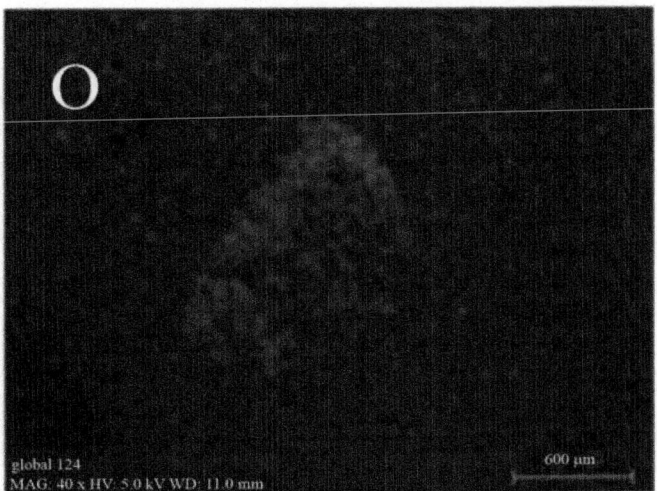

(c)

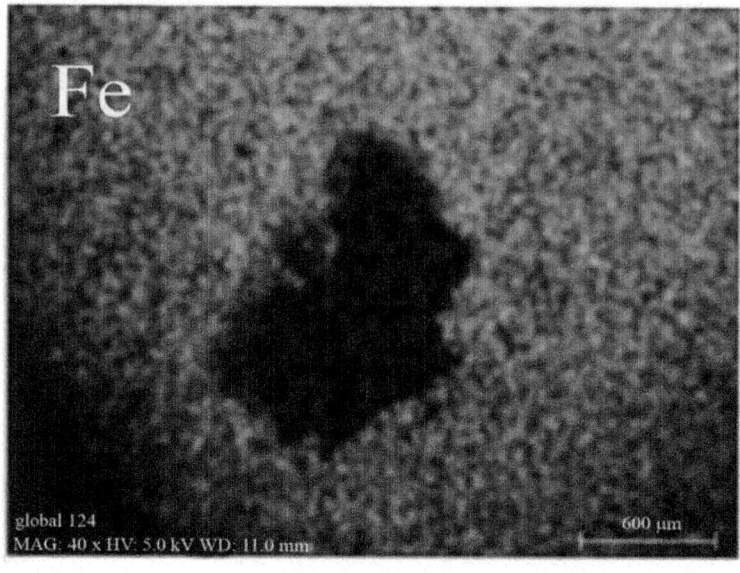

(d)

(e)

Figure 7: Left: SEM images of the 100Cr6 pin after sliding on the PA–Al composite indicating material transfer from the sample to the pin. Right: chemical maps of the pin obtained by wavelength-dispersive x-ray spectroscopy.

Wear Tests

Wear resistance of SLS samples was evaluated under lubricating water conditions using a standard polishing technique (SiC paper, 500 grit, normal load 10 N, 1 min, 150 rpm). Figure 8 compares the wear volumes measured for different composites produced by SLS or by hot pressing (180 °C, 5 min, 15 MPa). The QC filler particles provide the best wear resistance compared to all other types investigated, with a gain of up to 70% compared to unfilled PA or PA–Al composite. The comparison between SLS and hot-pressed samples shows similar properties, indicating that the composites are not affected by the SLS process compared to other conventional polymer manufacturing technologies.

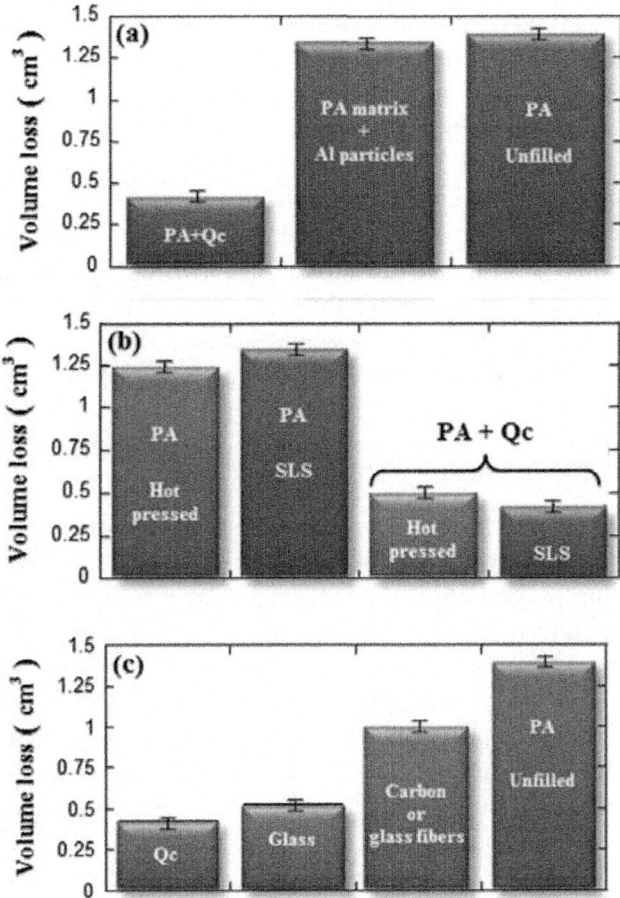

Figure 8: (a) Volume loss for SLS samples. When aluminum particles are added to the PA matrix, no gain is observed compared to unfilled PA. Quasicrystalline particles

(QCs) reduce the volume loss by about 70%. (b) Volume loss results of hot pressed samples compared to SLS samples. (c) Comparison of SLS composites reinforced by QC particles with commercial SLS samples filled by glass particles, carbon fibers and glass fibers. Adapted from [14], Copyright (2012), with permission from Elsevier.

Sealing Properties

Surprisingly, and in contrast to conventional SLS composites, the relative density of the composite parts reinforced by QC particles is close to 99% of the theoretical density. This value only reaches 85% for composite reinforced by Al particles. This very low porosity makes the PA–QC composites directly leak tight under air or water pressure (up to 7 bars) and for operating temperatures larger than 100 °C. This sealing property is only met for PA–QC composites. For all other commercial composite powders with similar particle size and volume fraction, an additional step of post-impregnation of the parts with a resin is necessary in order to fill the residual porosities and make the parts leak tight. This additional step increases the manufacturing time of the parts by approximately a factor of 2. Therefore the sealing property of PA–QC parts is a great advantage for many fluidic applications. The functional part in figure 4 is one such example. It shows an intake manifold of a car engine with wall thickness of only 2 mm. The fact that the new composite material is directly leak tight allows to one fabricate this part in one single operation, whereas using conventional materials, one would have to first build the parts in two halves by SLS, then process with resin impregnation and finally bond the two halves together. Other examples already on the market include hoses, inserts, adaptors, connectors, etc used in automotive engineering but other niche applications emerge as well like domestic appliances.

The origin of this sealing property must relate to some specific properties of the QC particles compared to other filler materials. One hypothesis could be that the wetting properties of the melted polymer are better on QC particles, allowing for reduced residual porosity of the SLS parts. We have tested this hypothesis by measuring contact angles formed by liquid polymer droplets deposited on a bulk sample of either the QC phase or a pure Al sample prepared by sintering. However, a good wettability (contact angle lower than 90°) is found in both cases. Another hypothesis would be that the effect of the laser beam during the sintering process is more effective in melting the PA powders in the presence of QC particles compared to Al particles, such that the polymer binder remains liquid for a longer time upon beam exposure thus leaving less porosity. As mentioned earlier in the introduction, the powder bed in SLS is pre-heated at a temperature that is just a few degrees below the melting point of the polymer binder and the laser beam only provides locally the additional energy required to melt the binder. The optical properties of QCs are known to differ significantly from normal metals like Al, absorbing more efficiently the IR light of the CO_2 laser beam. Therefore one expects that the embedded QC particles will adsorb more efficiently the laser beam energy, leading to a larger temperature increase and more effective melting of the surrounding polymer. In addition, the thermal conductivity of Al–Cu–Fe–B QC is about one hundred times lower than that of Al, a property which should further contribute to a more effective melting upon beam exposure.

QUASICRYSTAL-RESIN COMPOSITES FOR STEREOLITHOGRAPHY

SL is another additive manufacturing technology [24]. Its principle is similar to the SLS technology in the sense that it is also used to build 3D objects layer-by-layer. The main difference is that it uses a UV laser beam to polymerize locally a liquid resin monomer according to the 2D pattern of the CAD model, whereas SLS uses an IR laser to melt locally the polymer binder. Physically, the solidification consists in a photo-initiated polymerization, linking monomers into larger molecules by free radical reaction upon laser exposure. The resin must be solidified to a depth that must be larger than the layer thickness such that one layer adheres to the previous one. Once a layer has been completed, the part support is lowered into the vat containing the liquid resin by one layer thickness. The thin layer of liquid resin is then solidified (polymerized) according to the next 2D section and the process is repeated until the 3D part is completed, from bottom to top. The SL technology is widely used for both prototyping and manufacturing of functional parts for end users. However one disadvantage of this method compared to other rapid fabrication technologies is the limited range of available materials. The photo-curable resin can be filled with ceramic powders, and also glass or carbon fibers. Composites reinforced by metal particles have not been developed except for very small particle size, making its cost prohibitive for practical applications. The technological bottleneck is that filler particles added to the resin act as scattering centers for the incident light beam because of partial reflection and diffusion at the particle surfaces, limiting the penetration depth of the UV beam and thus the layer thickness.

Here again, the optical properties of QCs have been decisive in circumventing this problem and a QC–resin mixture could be successfully developed to build 3D functional parts by SL. The photo-curable resin is a commercial epoxy (Accura Si40®, 3D Systems). It contains cycloaliphatic epoxy resin (30–60 wt.%), aliphatic polyolpolyglycidyl (5–20 wt.%) and photoinitiators (0.1–10 wt.%) and has a density of 1.1 g cm^{-3}. The same Al–Cu–Fe–B QC particles produced by gas atomization were used as reinforcement particles. They were sieved with a mesh size of 25 μm. The optical reflectivity of the QC powders has been measured in the UV–visible range and compared to that of Al raw powders with similar particle size distribution. It was found that the reflectivity of the QC powder is much lower than that of Al at the wavelength of the UV laser used in SL (UV Nd–YAG laser, 355 nm wavelength).

The resin and the QC particles were mixed using a planetary mill with ceramic balls for 40 min at 100 rpm and ball-to-mixture weight ratio of 2:1. The optical reflectivity of the resin mixed with 20% volume of QC particles has been measured and compared to unfilled resin or resin mixed with 20% volume of Al particles as well as to a commercial resin filled with nanosized ceramic particles named Bluestone [25] (figure 9). The reflectivity is below 5% at the wavelength of the UV laser for the unfilled resin, the commercial Bluestone resin and the QC–resin mixture. It becomes much larger (15%) when Al particles are used, indicating that the photo-polymerization of the mixture will be less efficient in this case. The polymerization depth of the QC–resin mixture was measured as a function of the density of energy provided by the laser and for different volume fraction of the filler particles. The largest polymerization depth (70 μm) was obtained for a mixture containing 20% volume of QC particles and a laser energy density of 2 J m^{-2}. These values are compatible with the fabrication of 3D objects by adding up layers of about 50 μm in thickness, thus providing sufficient overlap between two consecutive layers. Note that under similar experimental conditions, the polymerization depth measured for the same resin filled with 20% volume of Al particles was lower than 30 μm, making it impossible to consider such composite for manufacturing by SL. Therefore the unusual optical properties of QC materials compared to normal metals [26] are a definite advantage when considering these materials as filler particles in SL resins.

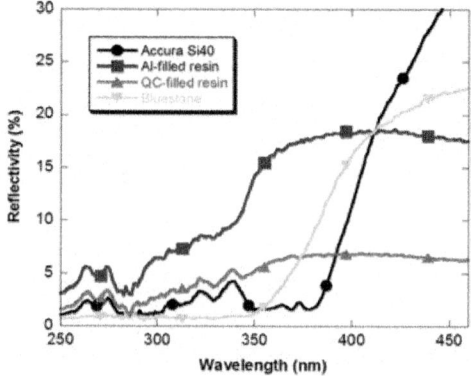

Figure 9: Reflectivity measurements in the UV–visible region of the different samples surfaces. Adapted from [15], Copyright (2014), with permission from Elsevier.

An example of such QC–resin 3D parts obtained by SL is shown in figure 10. Cross-sections of the parts have been investigated by optical microscopy revealing a homogeneous distribution of the QC particles, both within a layer and in the direction perpendicular to the layers. A linear shrinkage of the parts is typically observed in SL, due to the volume change of the polymer resin upon solidification. For a 25 mm diameter cylinder made of unfilled resin, the linear shrinkage is estimated to about 1.9% and is reduced to ~1.3% for the commercial Bluestone resin and to ~1% for the QC filled resin (20% vol.), allowing for improved manufacturing precision.

Figure 10: Picture of QC–resin composite parts produced by stereolithography. Adapted from [15], Copyright (2014), with permission from Elsevier.

Finally, friction and wear behavior of QC–resin composite parts have been tested and compared to that of the unfilled resin as well as of the Bluestone resin which presents the best performance available on SL market. The Shore D hardness of the QC–resin composites (87.5 ± 1) is found to be intermediate between the unfilled resin (83 ± 1) and the Bluestone resin (92 ± 1). The wear resistance was evaluated under lubricating water conditions using a standard polishing technique (SiC paper, 500 grit, normal load 10 N, 30 s, 150 rpm). The wear losses are reduced by ~40% in the case of QC filled resin compared to the unfilled resin, close to the value observed for the Bluestone nanocomposites (figure 11). The friction coefficients were measured using a pin-on-disk tribometer at room temperature in ambient atmosphere with a relative humidity of 50–60% and under non-lubricated conditions. The indenter was a 6 mm ball of 100Cr6 hard steel. The radius of the trace was 5 mm. The sliding velocity was 7.85 cm s^{-1} (150 rpm) and the normal load was 5 N. The results are shown in figure 11 as a function of the sliding distance.

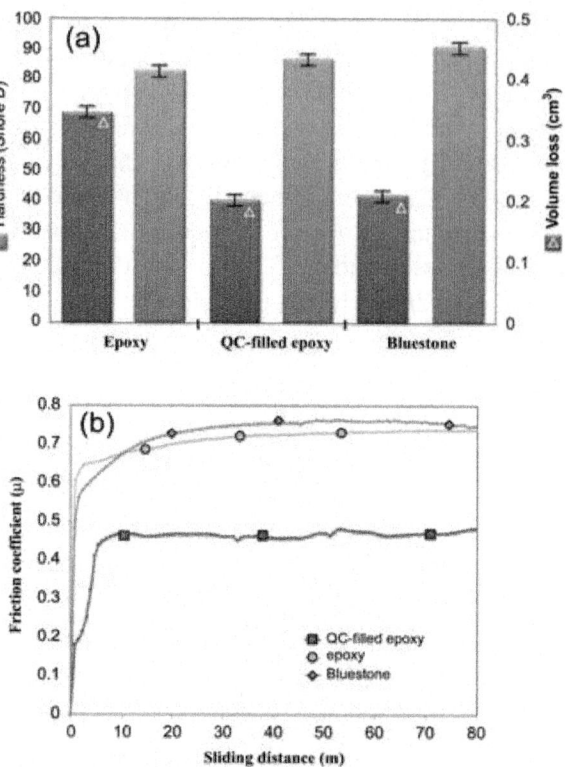

Figure 11: (a) Shore D hardness values (left scale) and abrasive volume loss (right scale) measured for the pure resin, the QC–resin composite and the commercial Bluestone resin. (b) Friction coefficients as a function of the sliding distance measured for the same samples. Adapted from [15], Copyright (2014), with permission from Elsevier.

The friction coefficients show the same behavior for the unfilled resin and the Bluestone, with a value of ~0.7. The friction coefficient becomes unstable in the case of the unfilled resin after 80 m of sliding distance as a result of material damage in the contact area. The friction forces are reduced by ~30% in the case of the QC filled resin and remain stable. These results are similar to those reported in section 4.1 for PA matrix reinforced by QC particles made by SLS. In addition, a pronounced abrasion of the 100Cr6 pin was observed by SEM after the friction test on the Bluestone composite and chemical maps revealed some material transfer from the composite sample to the pin. This is due to the presence of silica nanoparticles in the Bluestone resin which are known as strongly abrasive. Some material transfer was also observed from the QC–resin composite to the pin (presence of aluminum oxide on the pin) but no significant abrasion was reported in this case. Therefore the QC–resin composite for SL manufacturing might prove useful for industrial applications requiring good mechanical properties and reduced friction coefficients [15, 18].

CONCLUSIONS

The field of additive manufacturing technologies is growing rapidly and concerns many industrial sectors such as automotive, aerospace, tooling, medical and dental, design, etc. These technologies are conceptually similar to 3D printing and are not limited to polymer materials. They allow the manufacture of series of parts with more and more demanding functional properties and therefore the development of new materials adapted to these technologies is important.

In this article, we have demonstrated that CMAs such as QCs can be useful to design new composite materials adapted to additive manufacturing technologies. Light-weight metal–matrix composites have been developed to build 3D parts having mechanical properties similar to those of steel–brass composites used currently in the industry but having a density lower by a factor of 2. Polymer matrix composites reinforced by QC particles have been developed for both the direct SLS and SL technologies. These composites present several advantages in comparison to other materials currently available on the market and some of them are now commercialized [17, 27].

ACKNOWLEDGMENTS

This work was supported by the National Centre for Scientific Research (CNRS), the Region Lorraine and Ateliers CINI SA.

REFERENCES

1.   Wohlers T 2013 Wohlers Report, Additive Manufacturing and 3D Printing State of the Industry (Colorado: Wohlers Associates)

2.   Deckard C, Beaman J J and Darrah J 1988 Patent Cooperation Treaty Application WO9208567

3.   Kruth J P, Vandenbroucke B, Van Vaerenbergh J and Mercelis P 2005 Proc. 1st Int. Conf. on Polymers and Moulds Innovations (Belgium) 525

4.   Stewart T D, Dalgarno K W and Childs T H C 1999 Mater. Des. 20 133

5.   Sercombe T B and Schaffer G B 2003 Science 301 1225

6.   Mazzoli A, Moriconi G and Pauri M G 2007 Mater. Des. 28 993

7.   Shechtman D, Blech I, Gratias D and Cahn J W 1984 Phys. Rev. Lett. 53 1951

8. Dubois J M 2012 Chem. Soc. Rev. 41 6760

9. Dubois J M, Kang S S and Massiani Y 1993 J. Non-Cryst. Solids 153-154 443

10. Tsai A P, Aoki K, Inoue A and Masumoto T 1993 J. Mater. Res. 8 5

11. Inoue A and Kimura H 2000 Mater. Sci. Eng. A 286 1

12. Bloom P D, Baikerikar K G, Otaigbe J U and Sheares V V 2000 Mater. Sci. Eng. A 294-296 156

13. Sordelet D J, Besser M F and Logsdon J L 1998 Mater. Sci. Eng. A 255 54

14. Kenzari S, Bonina D, Dubois J M and Fournée V 2012 Mater. Des. 35 691

15. Sakly A, Kenzari S, Bonina D, Corbel S and Fournée V 2014 Mater. Des. 56 280

16. Kenzari S and Fournée V 2008 French Patent FR 2929541 (2009 Patent Cooperation Treaty Application WO2009/144405)

17. Kenzari S and Fournée V 2009 French Patent FR 2950826 (2011 Patent Cooperation Treaty Application WO2011/039469)

18. Sakly A, Kenzari S, Bonina D, Corbel S and Fournée V 2012 French Patent FR 2990375 (2013 Patent Cooperation Treaty Application WO2013/167448)

19. Sercombe T B and Schaffer G B 2004 Acta Mat. 52 3019

20. Lee S M, Jung J H, Fleury E, Kim W T and Kim D H 2000 Mat. Sci. Eng. A 294-296 99

21. Tang F, Anderson I E and Biner S B 2003 Mat. Sci. Eng. A 363 20

22. Sheares V V and Bloom P D 2000 Patent Cooperation Treaty Application WO2000/056538

23. Bloom P D, Baikerikar K G, Anderegg J W and Sheares V V 2001 643 K1631–12

24. André J C, Le Mehaute A and De Witte O 1984 French Patent FR 8411241

25. 3D Systems Accura® Bluestone material datasheet (www.3dsystems. com, accessed April 2014)

26. Demange V, Milandri A, de Weerd M C, Machizaud F, Geandel G and Dubois J M 2002 Phys. Rev. B 65 144205

27. Kenzari S and Fournée V 2011 French Patent FR 2979269 (2013 Patent Cooperation Treaty Application WO 2013/026972)

Chapter 7

A BILAYER RESOURCE MODEL FOR CLOUD MANUFACTURING SERVICES

Linan Zhu[1,2,3], Yanwei Zhao[1], and Wanliang Wang[2]

[1]Key Laboratory of Special Purpose Equipment and Advanced Processing Technology, Ministry of Education, Zhejiang University of Technology, Hangzhou 310014, China

[2]School of Computer Science and Technology, Zhejiang University of Technology, Hangzhou 310023, China

[3]College of Educational Science and Technology, Zhejiang University of Technology, Hangzhou 310023, China

ABSTRACT

Cloud Manufacturing and Cloud Service is currently one of the main directions of development in the manufacturing industry. Under the Cloud Manufacturing environment, the characteristics of publishing, updating, searching, and accessing manufacturing resources are massive, complex, heterogeneous, and so forth. A bilayer manufacturing resource model with separation of Cloud End and Cloud Manufacturing Platform is proposed in this paper. In Cloud End, manufacturing resources are divided into single resource and complex resource, and a basic data model of manufacturing resources oriented to enterprise interior is established to store the physical characteristics. In Cloud Manufacturing Platform, a resource service attribute model oriented to actual users is established to store the service characteristics. This model is described in detail and realized with stateful Web Service Description Language (WSDL) document. An example is provided for illustrating the implementation of the concept.

INTRODUCTION

The development and transformation of manufacturing has promoted the sustainable development of human society. At present, along with the generation and development of computer science and Internet, networked manufacturing, which contains production of material products and offering of immaterial services or functionality, has become the main form of manufacturing [1]. Attributed to Cloud Computing theory and application, Cloud Manufacturing and Cloud Service have gradually risen and become the main direction of manufacturing industry.

Presently, there is not a standardized definition of Cloud Manufacturing in academia. Li et al. believe that Cloud Manufacturing is a new networked manufacturing model that provides users with customized manufacturing services by organizing online manufacturing resource (called resource cloud) with the use of Internet and Cloud Manufacturing Service Platform [2]. They also put forward that Cloud Manufacturing is a service-oriented, high efficiency and low consumption, networked, and agile manufacturing model and technology. It enriches and expands Cloud Computing in two aspects of shared resource contents and service models, so it makes the manufacturing model become more agile, servicesation, environment friendly and intelligence [3]. According to Yang's point of view (China Aerospace Science and Technology Corporation) [4], the advantage of Cloud Manufacturing is that we can expand the philosophy of "Software as a Service" to "Manufacturing as a Service," so as to offer products with such services as high value-added, low cost, and global manufacturing under network environment. The authors of [5] believe that, by integration with contemporary technologies such as manufacturing informatization, Cloud Computing, Internet of Things, semantic web, and high-performance computing, Cloud Manufacturing expands and innovates the existing technology of networked manufacturing and service, and makes manufacturing resources and manufacturing capabilities become more virtualization and servicesation. Thus we can manage and operate the resources and service unified, centralized, and intelligently, to provide available, on-demand, reliable, and high-quality and low-cost services in every phase of manufacturing lifecycle. Cloud Manufacturing reflects the idea of "distributed resources are integrated to be used for one task" and "integrated resources are distributed to be used," and achieves the many-to-many service model, which provides multiple users with services at the same time by aggregating and centralized managing of distributed resources and services. From the perspective of product life cycle, the authors of [6] believe that Cloud Manufacturing, as a kind of manufacturing service, has its own lifecycle with several phases: the definition of manufacturing resource or capability, the provision of manufacturing resource or capability, the ordering of manufacturing task, manufacturing and distribution, and the disposal of manufacturing task.

Along with service-oriented technology, virtualization technology, Cloud Computing, and Internet of Things [7], the rapid development of manufacturing gives birth to a new model called Cloud Manufacturing Service Model (CMSM), which is a kind of service-oriented, high efficiency and low consumption, and knowledge-based networked intelligent manufacturing [2]. The CMSM has been actually accepted by some enterprises worldwide. In 2000 of America, MFG.com provided global manufacturing enterprises with an efficient trading platform, and the total value of average daily inquiry was more than 400 million yuan. Also in America, thanks to a brand new operation model, a micro-factory with the size of dry cleaner Local-Motors. com created a distinctive concept car called Rally Fighter in only 20 months. This is such a short time just equivalent to the time Detroit need to adjust the technical specifications of car door, because the total design tasks of the car were outsourced to community online, and the only task of Local-Motors. com was assembling parts, most of which were available on the market. In 2010 of China, a website for cutlery was built, which was a service system to help enterprises carry out cutlery sale and customer management better. The users of the website are the manufacturing enterprises in the supply chain or manufacturing chain of cutlery industry. Because of the website, the resource allocation involved in many cutlery enterprise supply chain and manufacturing chain is optimized.

The so-called Cloud Service is namely Resource Cloud Service, with some features: huge amounts of data, on-demand scaling, high availability, self-service interface, and flexible use of resources. Through the technologies such as Internet of Things and virtualization, Cloud Service virtually encapsulates manufacturing resources and manufacturing capabilities in different Cloud End based on knowledge, and intelligently accesses to Cloud Manufacturing Platform, thus provide users with manufacturing resources highly virtualized in Cloud End as services in the manufacturing life cycle [8].

CLOUD MANUFACTURING SYSTEM STRUCTURE

Showed as Figure 1, the Cloud Manufacturing System (CMS) is composed of Cloud End (CE) and Cloud Manufacturing Platform (CMP), and the CE contains Cloud Provider (CP) and Cloud Demander (CD). The CP and CD are, respectively, the providers and the demanders or users of resource cloud or cloud service. The CP provides corresponding manufacturing resource and service through the CMP; the CD proposes manufacturing demands and gets corresponding resources or services through CMP; according to the CD, the CMP searches suitable resources or services, and provides the CD with demanded resources or services by the use of Cloud technology, Cloud

Services management technology, Cloud Manufacturing security technology, and Cloud Manufacturing business management model and technology [2]. The CMP is also composed of many sub-platforms, which can communicate with each other to realize high sharing of resources. Also, the CMP embodies many modules such as resource database and middleware, which have powerful functions of resource scheduling.

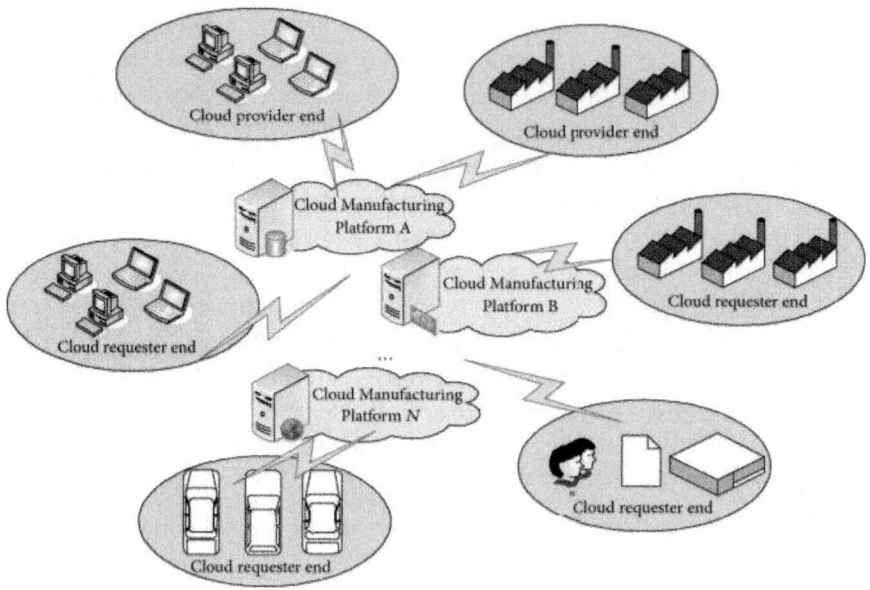

Figure 1: The system of Cloud Manufacturing.

NETWORKED MANUFACTURING RESOURCE

Resource description is distinctive in different manufacturing areas and with different modeling aims [9]. Cloud Manufacturing is actually a kind of complex network environment with the characteristics of large-scale, heterogeneous, and high sharing of resources. Under the network environment, the key points of manufacturing resource description are dynamic expression of resource service capabilities, rapid search of resources, optimal allocation of resource, and dynamic planning of the product life cycle.

Overview on Networked Manufacturing Resource Modeling

In [10], a notion named service domain was presented to help UML analyze and model manufacturing resource in manufacturing grid, and supported hierarchy management of resource model, and the results resolved the disadvantages that modeled huge manufacturing resource in manufacturing grid through only a set of UML class graph. In [11], the authors studied on the multi-dimension manufacturing resource modeling technology: the manufacturing resource life cycle dimension, manufacturing resource application view dimension, and polymerization granularity dimension, and established a multi-dimensional network ontology model of manufacturing resources through semantic web technology. In [12], a combined method of particle size based on ontology and business-oriented needs of internal and external information is proposed. The authors also proposed a model based on Web Process Planning, and expanded the OWL-S for describing the dynamic characteristics of the service. So the information model was set for discovering and matching the manufacturing resource. Aiming at the problem of manufacturing resource functionality similarity under nonlinear process planning environments, meta-resource methodology was introduced in [13], which could accurately express diversity of manufacturing resources and improve the agility of manufacturing system. For distributed networked manufacturing resources, a networked manufacturing resource model based on physical manufacturing unit, as well as an information model and an information integration method of manufacturing resource based on XML, were presented in [14], with the result of realization of heterogeneous manufacturing resource description. In order to implement the integration and application of legacy hardware/software manufacturing resources in manufacturing enterprises, a scheme and a framework for the manufacturing resource encapsulation and integration based on mobile agent was introduced in [15]. So the function of reconfiguration and encapsulation for legacy manufacturing resources, as well as information interaction and acquirement, could be realized.

Classification of Manufacturing Resources

With different manufacturing aims and different manufacturing methods, the classification of resource is different. In [12], the authors established resource ontology tree, and divided the resource into six categories: (1) financial resources: the financial elements of business operations including fixed assets, liquidity, and liabilities; (2) technical resources: process technology, industry standards, and so forth; (3) equipment resources: machine tools, tooling, gages, and so forth; (4) human resources: engineering and technical personnel, workers, technicians, management personnel, and so forth; (5) software resources: CAD software, financial software, ERP software, and so forth; (6) logistics resources: truck, train, and so forth. According to the sharing method of manufacturing resource under manufacturing grid environment, the authors of [16] divided resources into six categories: hardware resources, software

+ hardware resources, software resources, online resources, semionline resources, and offline resources.

In this paper, according to the properties of the resource, the users' needs, usage as well as the role played in manufacturing activities, we divide resources into eight categories (shown in Figure 2): human resources, manufacturing equipment resources, software resources, service resources, material resources, computing resources, manufacturing knowledge resources, and other resources. Human resource means the staff involved in all phases of product life cycle; manufacturing equipment resource means all kinds of hardware in the product design and production processing; software resource means computer software used in product design, manufacturing, enterprise management, and so forth; service resource means service activities related to product design, production marketing, and so forth; material resource means raw materials used in every phases of production; computing resource means computing equipment such as CPUs and memories used in production and enterprise management; knowledge resource means the knowledge or technology related to production.

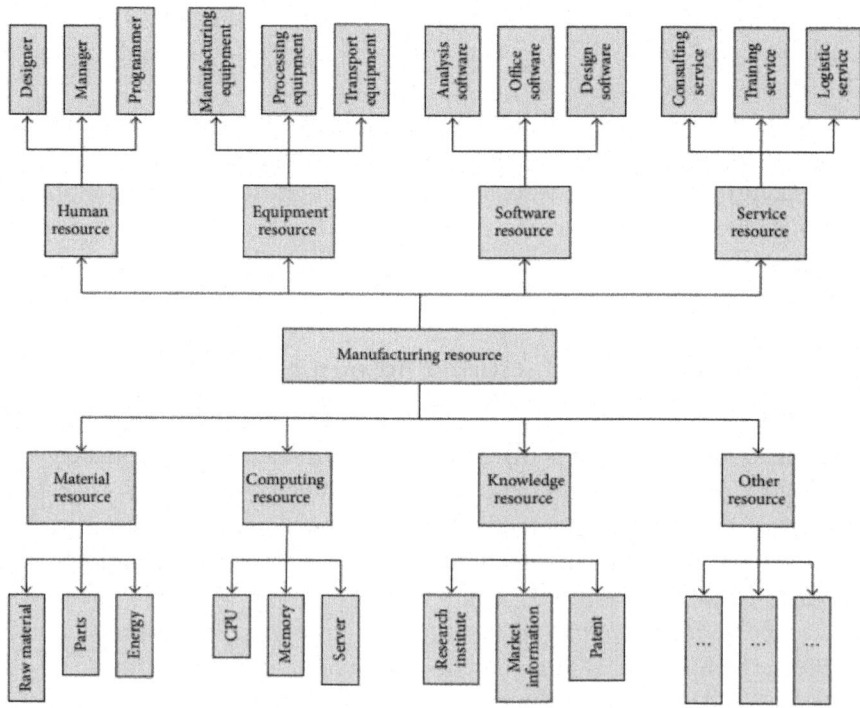

Figure 2: Manufacturing resource classification.

THE BILAYER RESOURCE MODEL

Under Cloud Manufacturing environment, there are a wide range of manufacturing resources in different areas, and the systems in different manufacturing units are heterogeneous. So the resource data has the characteristics of massive, complex and heterogeneous, and there is lack of uniform data standards. For the above reasons, it is difficult to real-time monitor, operate, or update resource data. Aiming at these questions, we propose a bilayer resource model with separation of CE and CMP (shown in Figure 3). In CE, we establish the resource basic data model, which is used to store the basic data of resources such as performance parameters, physical structure, input data type, output data type, geographic information, and manufacturer information, and provides access interface to CMP. In CMP, we establish the resource service attribute model, which is used to store service information such as function parameters, input data template, output structure template, and service quality evaluation. In CMP, there are also some resource service solutions, integrated resource optimization tools, and database for storing data accessed frequently according to usage history.

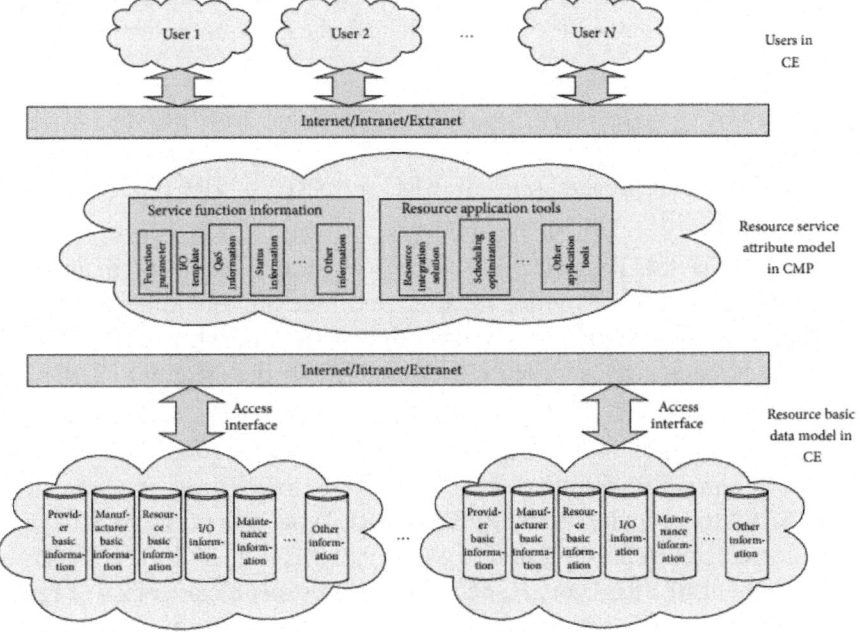

Figure 3: The bilayer resource model in Cloud Manufacturing.

Cloud End Model

As described above, resource model in CE is used to store related data of the actual manufacturing resource, which consists of single resource (SR) and complex resource (CR). According to the functionality, the SR is divided into eight categories, as described in Section 3.2; as a resource group, the CR is formed by combination of SR based on functionality and has some function. So, the formalized representation of resource cloud in CE is Re sCloudEnd = {RCEInfo, RC}.

RCEInfo means the basic information of the CE, and its formalized representation is *RCEInfo* = {*RCEID, RCE Provider, RCELocation, RCEOther*}. *RCEID* no longer changed after registration is the flag of CE, and it's used to uniquely identify CE so as to realize locating and indexing the manufacturing resource providers; *RCE Provider* is the name and contact information of CE, stored as a string; *RCELocation* is the location of CE, stored as a string; *RCEOther* is the custom information, which can be added or removed according to the actual situation, and the format of data is user-defined, usually as a string.

RC is resource cloud, and its formalized representation is: $R ::= (SRC, CRC)$. *SRC* is single resource cloud (SRC), and *CRC* is complex resource cloud (CRC).

Single Resource Cloud (SRC)

The formalized representation of SRC is SRC = {RCID, SRCBaseInfo, SRCFuncInfo, SRCStatusInfo}.

1. *RCID* is the flag of SRC, used to uniquely identify manufacturing resource cloud to realize locating and indexing manufacturing resource cloud. The uniqueness is within the same CE, while the same kind of manufacturing resource in different CE has the different RCIDs.

2. *SRCBaseInfo* is the basic characteristics set, and its formalized representation is *SRCBaseInfo* = {SRCName, SRCModel, SRCType, SRCInfo, SRCParameter, SRCBInfos}. *SRCName* is name of SRC; SRCModel is the model of SRC; SRCType is the type of SRC, and the formalized representation is $SRCType = \{1, 2, 3, 4, 5, 6, 7, 8\}$, every element of SRCType, respectively, represents 8 categories of resources as described in Section 3.2; *SRCInfo* is the description of SRC, stored as a string; *SRCParameter* is the parameters set of SRC, including size, and cost, defined by the provider according to specification of resources; *SRCBInfos* is custom information, which can be added or

removed according to the actual situation, and the format of data is user-defined, usually as a string.

3. *SRCFuncInfo* is the functionality characteristics attributes set of SRC, and its formalized representation is: SRCFuncInfo = {SRCTaskType, SRCCapa, SRCTime, SRCQua, SRCEnviro, SRCFInfos}. *SRCTaskType* is the service type supported by the SRC, such as the type of processing object; *SRCCapa* is the indicators of the processing object, such as input set of object, output set of object, and precision, and its formalized representation is: $SRCCapa = \{IObjSet, OObjSet, \ldots, Capa_n\}$, every element of *SRCCapa* has its own independent formalized representation. For example, $IObjSet = \{IObj_i \mid i = 1, 2, \ldots, n\}$, $OObjSet = \{OObj_i \mid i = 1, 2, \ldots, n\}$, every $IObj_i$ or $OObj_i$ is composed of some characteristic attributes: $IObj_i = \{ID, Name, P_1, P_2,...,P_n\}$, $OObj_i = \{ID, Name, P_1, P_2,...,P_n\}$; *SRCTime* is the minimum time and maximum time to complete the task, and its formalized representation is: $SRCTime = \{STime_{min}, STime_{max}\}$; *SRCQua* is the quality standards the resource can reach, and its formalized representation is: $SRCQua = \{N, level_1, level_2, level_3, \ldots, level_n\}$. N represents the sum of quality standards the offering can reach, $leveli$ store the name of quality standards as a string. Take the safety valve as an instance, the formalized representation is: $SRCQua = \{1, "GB150.1 - 2011"\}$; *SRCEnviro* is the service environment requirements of the SRC, such as geographical requirements, software platform requirements, and enterprise-class requirements; *SRCFInfos* is custom information, which can be added or removed according to the actual situation, and the format of data is user-defined, usually as a string.

4. *SRCStatusInfo* is the status set of SRC, and its formalized representation is: $SRCStatusInfo = \{0, 1, 2, 3, 4, 5\}$. 0 represents that the resource is in maintenance; 1 is idle; 2 is under load; 3 is full load; 4 is over load; 5 is invalid.

Complex Resource Cloud (CRC)

The formalized representation of CRC is: CRC = {RCID, SRCSet, SRCRelSet, CRCFuncInfo, CRCStatusInfo}.

1. *RCID* is the flag of CRC, and has the same characteristics with the flag of SRC.

2. SRCSet is the SRCs set contained by the CRC, and its formalized representation is: $SRCSet = \{N, SRC_1, SRC_2, \ldots, SRC_n\}$. N is the sum of SRCs, and SRC_i is the flag *RCID* of SRC.

3. SRCRelSet is the logical relationships between these SRCs, and its formalization representation is: SRCRelSet ::= {Rand, Ror, Rprev, Rnext}. Rand is the parallel relationship. Rand(A, B) means that the resource A and B have no influence with each other, and can be used in parallel; Ror is the selection relationship. $Ro(A, B)$ means that the resource A and B have the same or similar functionalities, and are alternative in the actual manufacturing; Rprev and Rnext are the precursor relationship and successor relationship. $Rpre(A, B)$ means that A is the precursor of B, and $Rnext(A, B)$ means that A is the successor of B, so the relationship of resource A and B is serial. It's worth nothing that $Rpre(A, B) \neq Rnext(B, A)$, because the precursor relationship or successor relationship may be one to many or many to many.

4. $CRCFunc$Info is functionality characteristics attributes set, and its formalized representation is: CRCFuncInfo = {CRCTaskType, CRCCapa, CRCTime, CRCQua, CRCEviro, CRCFInfos}. $CRCTaskType$, $CRCCapa$, $CRCQua$, $CRCEviro$ and $CRCFInfos$ respectively means supported service type, indicator requirements, quality standards, environment requirements and custom contents, with the same meaning and formalized representation with $SRCTaskType$, $SRCCapa$, $SRCQua$, $SRCEviro$ and $SRCFInfos$of SRC. $CRCTime$ is the minimum time and maximum time to complete the task, and its formalized representation is $CRCTime = \{CTime_{min}, CTime_{max}\}$. $CTime_{min}$ is the sum of $STime$min of all serial single resource units (the formula is shown as (1)), taking the minimum value of several parallel single resource (the formula is shown as (2)); $CTime_{max}$ is the sum of $STime_{max}$ of all serial single resource units (the formula is shown as (3)), taking the maximum value of several parallel single resource (the formula is shown as (4)):

$$CTime_{min} = SRC_1 \cdot STime_{min}$$
$$+ SRC_2 \cdot STime_{min} + \cdots$$
$$+ SRC_{i1,i2...,ij} \cdot STime_{min} + \cdots$$
$$+ SRC_n \cdot STime_{min} \tag{1}$$

$$SRC_{i1,i2...,ij} \cdot STime_{min} = min \{ SRC_{i1} \cdot STime_{min},$$
$$SRC_{i2} \cdot STime_{min}, \ldots,$$
$$SRC_{ij} \cdot STime_{min} \} \tag{2}$$

$$CTime_{max} = SRC_1 \cdot STime_{max}$$
$$+ SRC_2 \cdot STime_{max} + \cdots$$
$$+ SRC_{i1,i2\ldots,ij} \cdot STime_{max} + \cdots$$
$$+ SRC_n \cdot STime_{max} \qquad (3)$$

$$SRC_{i1,i2\ldots,ij} \cdot STime_{max} = \max \{ SRC_{i1} \cdot STime_{max},$$
$$SRC_{i2} \cdot STime_{max}, \ldots ,$$
$$SRC_{ij} \cdot STime_{max} \} . \qquad (4)$$

1. CRCStatusInfo is the status set of CRC, and its formalized representation is: CRCStatusInfo $= \{0, 1, 2, 3, 4, 5\}$. 0 represents that the resource is maintenance; 1 is idle; 2 is under load; 3 is full load; 4 is over load; 5 is invalid. Here, we transform SRCSet into SRCSetX, SRCSetX $= \{X, SRC_1, SRC_2, ..., SRC_{i1,i2}, ..., SRC_{ii,ii+1,\ldots,ij}, ..., SRC_x\}$. X is the sum of single resource unit, and $SRC_{i1,i2}$ and $SRC_{ii,ii+1,\ldots,ij}$ both represent a single resource unit composed of several parallel single resources, and the formula of single resource unit resource statues is (5); so the formula of CRCStatusInfo is (6):

$$SRC_{ii,ii+1,\ldots,ij} \cdot SRCStatusInfo$$

$$= \begin{cases} 0, & SRC_j = 0, j = ii, ii + 1, \ldots, jj \\ 1, & \exists SRC_j = 1, j = ii, ii + 1, \ldots, jj \\ 2, & \exists SRC_j = 2, j = ii, ii + 1, \ldots, jj \\ 3, & SRC_j = 3, j = ii, ii + 1, \ldots, jj \\ 4, & SRC_j = 4, j = ii, ii + 1, \ldots, jj \\ 5, & SRC_j = 5, j = ii, ii + 1, \ldots, jj \end{cases}$$

$$(5)$$

CRCStatusInfo

$$= \begin{cases} 0, & \exists SRC_i \cdot SRCStatusInfo = 0, \\ & SRC_i \in SRCSetX \\ 1, & SRC_i = 1, SRC_i \in SRCSetX \\ 2, & SRC_i = 2, SRC_i \in SRCSetX \\ 3, & SRC_i = 3, SRC_i \in SRCSetX \\ 4, & SRC_i = 4, SRC_i \in SRCSetX \\ 5, & \exists SRC_i \cdot SRCStatusInfo = 5, \\ & SRC_i \in SRCSetX. \end{cases}$$

(6)

Cloud Manufacturing Platform Model

The user is the main service object of CMP, so it's needed to establish a service-oriented resource model called resource service attribute model to provide users with information related to resource functionality, usage and service quality, all of which can help users search, select and use resource. The formalized representation of resource model in CMP is Re sCloudPlatform = {ID, RCEID, BaseInfo, ServiceInfo, AssessInfo, OtherInfo}.

1. *ID* is the flag of resource cloud, used to uniquely identify manufacturing resource cloud in the same subCMP so as to realize locating and indexing resource. The uniqueness is within the same sub-CMP, while the same *ID* in different sub-CMP may represent different resource cloud, thus avoiding updating the resource list of all the subCMPs once a CE publishes or updates resource to a certain sub-CMP. It is more suitable for the massive and complex characteristics of resource cloud data.

2. *RCEID* is the CE flag which provides the resource cloud, used to uniquely identify the provider of resource.

3. *BaseInfo* is the basic characteristics attributes set of resource cloud, and its formalized representation is *BaseInfo* = {*Name, Model*, Re *sInfo*, Provider}. *Name* is the resource name; *Model* is the resource model; Re *sInfo* is the description information of resource, defined by the provider; Provider is the description of resource provider, defined by provider-self.

4. *ServiceInfo* is the service capability information, and its formalized representation is ServiceInfo = {TaskType, Capa, Time, Qua, Cost, Enviro, photo, OtherServInfo}. The elements respectively mean service type, performance indicators, time-consuming, quality standards, price,

service environment, other service information, and so forth.

5. *AssessInfo* is the assessment information of resource, and its formalized representation is AssessInfo = {TaskID, TAss, CAss, QAss, ServAss, CreditAss, Average}. *TaskID* is the identification number of a task, automatically generated by sub-CMP; *TAss* is time-consuming evaluation; *CAss* is price evaluation; *QAss* is quality evaluation; *ServAss* is service evaluation; *CreditAss* is credit evaluation; *Average* is average value of every indicator up to now

6. *OtherInfo* is other service capability information, added or removed by resource provider, sub-CMPs and users, and its formalized representation is $OtherInfo = \{other \inf o_1, other \inf o_2, \ldots, other \inf o_i, \ldots\}$, $other \inf o_i = \{user_i, \inf o_i\}$, $user_i$ is the provider of other service capability information, $\inf o_i$ is the information content stored as a string.

REALIZATION

We build the Cloud Manufacturing Resource Model with stateful Web Service Description Language (WSDL) document, when dynamically publishing and updating resource cloud. Take certain vertical CNC machine tools as an instance, its processing service capability can be simply described as Table 1, the WSDL document of single resource cloud in CE is shown as Algorithm 1, and the WSDL document of resource cloud in CMP is shown as Algorithm 2.

Table 1: Parameters of vertical CNC machine tools

Attribute	Value	Attribute	Value
ID	CC615	Crosshead trip	760 mm
Name	C5116E Machine	Turret fast moving speed	1800 mm/min
Introduction	......	Vertical turret slewing angle	30±°
Enterprise	......Ltd.	Vertical turret maximum cutting force	25 KN
Address	No. 123 Liuhe Road, Hangzhou, Zhejiang Province	Side turret maximum cutting force	20 KN
Price	28 yuan/hour	Workbench maximum torque	25 KNM
Maximum turning diameter	1600 mm	Tool bar section height	40 mm
Table diameter	1400 mm	Main motor power	22/30 KW
Maximum processing height	1000 mm	Overall dimensions (length × width × height)	2800 × 3000 × 3200 mm³
Maximum workpiece weight	5000 kg	Machine weight	12580 kg
Table speed range (16)	5–160 r/min	Photo gallery	CC615_1.jpg CC615_2.jpg CC615_3.jpg
Turret feed range (12)	0.8–86 mm	Years in use	N years
Vertical turret ram stroke	800 mm	Assess	{001, 1, 672, 4, 1, "comment001"}{002, 1, 336, 5, 2, "commen002"}{003, 0.5, 163, 3, 3, "commen003"}...... {000, 0.83, 4, 2,}
Side head ram level trip	630 mm	Status	{0, 1, 2, 3, 4, 5}
Side head vertical trip	980 mm	......	......
Side head maximum turning diameter	1400 mm	......	......

```
< ResCloudEnd >
    < ResCloudEnd-info indentifier = "single-resource" >
        < RCEInfo >
            < description name = "RCEID" value = "00888" />
            < description name = "RCEProvider" >
                < description name = "RCEProvider-name" value = "x x x Ltd." />
                < description name = "RCEProvider-tel" value = "86-0571-12345678" />
                ...
            </ description >
            < description name = "RCEAddress" value = "No. 123, Liuhe Road, Zhejiang Province" />
            < description name = "RCEOther" value = null />
        </ RCEInfo >
        < RC >
            < description name = "RCID" value = "CC615" />
            < description name = "SRCBaseInfo" >
                < description name = "SRCName" value = "Single Column Vertical Machine Tool" />
                < description name = "SRCModel" value = "C5116E" />
                < description name = "SRCType" value = "2" />
                < description name = "SRCInfo" value = null />
                < description name = "SRCParameter" >
                    < description name = "length" value = "2800" />
                    < description name = "Width" value = "3000" />
                    < description name = "height" value = "3200" />
                    < description name = "power" value = "22/33" />
                    ...
                </ description >
                < description name = "SRCBInfos" value = null />
            </ description >
            < description name = "SRCFuncInfo" >
                < description name = "SRCTaskType" value = null />
                < description name = "SRCCapa" >
                    ...
                    < description name = "workpiece weight" value = "5000 kg" />
                    ...
                </ description >
                ...
            </ description >
            < description name = "SRCStatusInfo" value = "2" >
        </ RC >
    </ ResCloudEnd-info >
</ ResCloudEnd >
```

Algorithm 1: WSDL of single resource cloud description in CE.

```
< ResCloudPlatform >
   < ResCloudPlatform-info indentifier = "ResCloudPlatform" >
      < description name = "ID" value = "SB0088" />
      < description name = "RCEID" value = "00888" />
      < description name = "BaseInfo" >
         < description name = "name" value = "Single Column Vertical Machine Tool" />
         < description name = "Model" value = "C5116E" />
         < description name = "ResInfo" value = null />
         < description name = "Provider" value = "× × × Ltd." />
      </ description >
      < description name = "ServiceInfo" >
         ...
         < description name = "Price" value = "28 yuan/hour" />
         < description name = "photo" >
            < description name = "photo1" >< a href = CC615_1.jpg > photo1 < /a >< / description >
            < description name = "photo2" >< a href = CC615_2.jpg > photo2 < /a >< / description >
            < description name = "photo3" >< a href = CC615_3.jpg > photo3 < /a >< / description >
            ...
         </ description >
         ...
      </ description >
      < description name = "AssessInfo" >
         ...
         < description name = "IncreditAss-Average" average = "2" >
            < description name = "IncreditAss-001" TaskID = "001" IncreditAss = "1" />
            < description name = "IncreditAss-002" TaskID = "002" IncreditAss = "2" />
            < description name = "IncreditAss-003" TaskID = "003" IncreditAss = "3" />
         </ description >
         ...
      </ description >
      < description name = "OtherInfo" >
         < description name = "otherinfo1" attribute1 = "John" attribute2 = "value1" />
         < description name = "otherinfo2" attribute2 = "Microsoft" attribute2 = "value2" />
         ...
      </ description >
   </ ResCloudPlatform-info >
</ ResCloudPlatform >
```

Algorithm 2: WSDL of resource description in CMP.

DISCUSSION

Cloud Manufacturing is a kind of intelligent networked manufacturing model with the characteristics such as service-oriented, high-efficiency and low-energy, and knowledge-based. Through the integration of some advanced technology [17–19], the manufacturing resources would be virtualization and servicesation, in order to be intelligently, multi-win-win, and efficiently shared and collaborated. In the Cloud Manufacturing environment, enterprises provide convenient, on-demand, safe and reliable, and high-quality and low-cost services. Cloud Manufacturing has the following characteristics: highly fragmented distribution and highly concentrated use of resources, service-oriented and demand-oriented, uncertainty of manufacturing plan, manufacturing with users involved, and use and pay on-demand. Because of all the above, the structure, functionality, operating environment, and basic physical attribute have the following characteristics: diversity, complexity, heterogeneity, being massive, and so forth. The bilayer model of manufacturing resource proposed in this paper can adapt well to such a manufacturing environment. Firstly, the separation of the basic data model and the service attribute model, between which a one-to-one mapping is formed by keywords, enables the bilayer model to adapt to the complex and heterogeneous resource in Cloud Manufacturing. Secondly, the two types of resource cloud (single resource cloud and complex resource cloud) can describe well the resource basic characteristics with the basic data model, which usually has a huge amount of data and is relatively certainty stored in CE, and thus greatly reduces heavy tasks for data storage, maintenance, and updating in CMP. thirdly, during dynamically publishing and updating resource cloud, we adopt the stateful Web Service Description Language (WSDL) document, which has a high level of flexibility, and thus is suitable for massive data easily accessed by CE in the Cloud Manufacturing environment. Finally, fully taking into account the user participation, user evaluation is included in the resource model, thus eases to updating and improving the manufacturing resources, and fully reflects the Cloud Manufacturing characteristics of user-oriented and service-oriented.

CONCLUSION

With the increasingly fierce competition in the global market and increasingly serious energy and environmental issues, how to provide users with high quality products and services with low energy consumption and environmental friendly, is an urgent problem to manufacturing industry currently [8]. Cloud Manufacturing and Cloud Service is one of the main directions of the current development in the manufacturing industry and results by applying Cloud Computing theory to manufacturing industry, so it has universal characteristics of "Cloud Theory" such as high user participation, high resource sharing, and high process agility. Aiming at making manufacturing agility, servicesation, green, and intelligence, Cloud Manufacturing is a new development of networked manufacturing with the service-oriented manufacturing theory called Manufacturing as a Service (MaaS) [20]. Under Cloud Manufacturing environment, the publishing, updating, searching, and accessing manufacturing resources have the characteristics of being massive, complexity, and heterogeneity, thus putting forward higher flexibility, agility, and versatility on the resource modeling. In this paper, a new bilayer manufacturing resource model with separation of CE and CMP is proposed. The model is comprised of resource basic data model and resource service attribute model: the resource basic data model in CE focuses on the physical characteristics of manufacturing resource, and it faces enterprise interior manufacturing management system. And the two type resource models (single resource and complex resource) can describe well the resource basic characteristics, which usually have a huge amount of data stored relatively certainly in CE; the resource service attribute model, which usually has smaller amount of data stored in CMP, focuses on service characteristics of manufacturing resource, and it faces resource users' actual demands. And fully taking into account the user participation, user evaluation is included in the resource model. Besides, the stateful WSDL documents can describe resource attributes perfectly and flexibly thus are suitable to the need under Cloud Manufacturing environment. Through realization of the instance, the model can run well in Cloud Manufacturing environment and can store, publish, and update large and complex data. The users' demands on discovering and selecting manufacturing resource and deeply participating in manufacturing process can also be met.

ACKNOWLEDGMENTS

The presented research was supported by the National Natural Science Foundation, China (no. 60970021), and the Scientific Research Plan of Zhejiang Provincial Department of Education, China (no. Y201120777).

REFERENCES

1. T. S. Baines, H. W. Lightfoot, S. Evans et al., "State-of-the-art in product-service systems," Proceedings of the Institution of Mechanical Engineers Part B, vol. 221, no. 10, pp. 1543–1552, 2007

2. B. H. Li, L. Zhang, S. L. Wang, F. Tao, and J. W. Cao, "Cloud manufacturing: a new service-oriented networked manufacturing model," Computer Integrated Manufacturing Systems, vol. 16, no. 1, pp. 1–7, 2011.

3. B. H. Li, L. Zhang, L. Ren et al., "Further discussion on cloud manufacturing," Computer Integrated Manufacturing Systems, vol. 17, no. 3, pp. 449–457, 2011.

4. H. W. Yang, Cloud Manufacturing: A Manufacturing Service, Manufacture Information Engineering of China, 2010.

5. B. H. Li, L. Zhang, and X. D. Chai, "Introduction to cloud manufacturing," ZTE Technology Journal, vol. 16, no. 4, pp. 5–8, 2010.

6. G. Breiter and M. Behrendt, "Life cycle and characteristics of services in the world of cloud computing," IBM Journal of Research and Development, vol. 53, no. 4, pp. 3:1–3:8, 2009, paper no. 3.

7.   K. Chen and W. M. Zheng, "Cloud computing: system instances and current research," Journal of Software, vol. 20, no. 5, pp. 1337–1348, 2009.

8.   F. Tao, L. Zhang, H. Guo, Y. L. Luo, and L. Ren, "Typical characteristics of cloud manufacturing and several key issues of cloud service composition," Computer Integrated Manufacturing Systems, vol. 17, no. 3, pp. 477–486, 2011.

9.   M. Lu, X. D. Sun, and G. Wang, "Resource modeling to support manufacturing process optimization under uncertain environment," Computer Integrated Manufacturing Systems, vol. 16, no. 12, pp. 2611–2616, 2010.

10.   B. Sheng, Y. Li, Y. Ding, F. Tao, and Z. Zhou, "Modeling and management of manufacturing resource information in manufacturing grid," China Mechanical Engineering, vol. 17, no. 13, pp. 1375–1380, 2006.

11.   Y. D. Fang, L. H. Du, W. P. He, H. Chen, and B. Sun, "Multi-dimension manufacturing resource modeling technology research based on networked manufacturing environment," Application Research of Computers, vol. 26, no. 2, pp. 559–562, 2009

12.   L. B. Zheng, J. N. Gu, and Y. R. Dai, "Modeling of manufacturing resources based on ontology,"Machine Design and Research, vol. 25, no. 5, pp. 61–63, 2009.

13.   F. Liang, Z. B. Jiang, and L. Y. Tao, "Model building approach of manufacturing resource based on meta-resource," Computer Integrated Manufacturing Systems, vol. 14, no. 12, pp. 2306–2311, 2008.

14.   C. F. Yao, D. H. Zhang, K. Bu, W. H. Wang, and J. X. Ren, "Networked manufacturing resources modeling and information integrationbased on physical manufacturing unit," Computer Integrated Manufacturing Systems, vol. 14, no. 4, pp. 667–674, 2008.

15.   G. H. Zhou and P. Y. Jiang, "Encapsulation and integration for networked manufacturing resources based on mobile agents," Computer Integrated Manufacturing Systems, vol. 8, no. 9, pp. 728–732, 2002.

16.   S. Y. Shi, H. B. Liu, H. C. Yang, R. Mo, and Z. F. Chen, "Resource modeling based on web service in manufacturing grid," Aeronautical Manufacturing Technology, vol. 12, pp. 80–83, 2008.

17.   S. Y. Chen, Y. F. Li, and N. M. Kwok, "Active vision in robotic systems: a survey of recent developments," The International Journal of Robotics Research, vol. 30, no. 11, pp. 1343–1377, 2011.

18. S. C. Lim, C. H. Eab, K. H. Mak, M. Li, and S. Y. Chen, "Solving linear coupled fractional differential equations by direct operational method and some applications," Mathematical Problems in Engineering, vol. 2012, Article ID 653939, 28 pages, 2012.

19. W. Huang and S. Y. Chen, "Epidemic metapopulation model with traffic routing in scale-free networks,"Journal of Statistical Mechanics: Theory and Experiment, vol. 2011, Article ID P12004, 19 pages, 2011.

20. L. Zhang, Y. L. Luo, W. H. Fan, F. Tao, and L. Ren, "Analyses of cloud manufacturing and related advanced manufacturing models," Computer Integrated Manufacturing Systems, vol. 17, no. 3, pp. 458–468, 2011

Chapter 8

IDENTIFYING ENABLERS OF E-MANUFACTURING

Rajeev Saha and Sandeep Grover

Department of Mechanical Engineering, YMCA University of Science and Technology, Sector-6, Faridabad 121006, Haryana, India

ABSTRACT

The term e-manufacturing refers to the ability of a manufacturing system to integrate various inputs using internet and intranet. With the market being increasingly competitive and customer oriented desiring to get best of the quality at cheapest available price of a product in shortest possible time, the role of e-manufacturing is pivotal towards the success of a company. Another important fact is that each customer may desire to have a certain different set of values added to the product being purchased. Customers want to voice their concerns directly to manufacturers thereby necessitating an interface to hear them in real-time and take suitable action thereupon, if needed, in real-time. As such identifying the enablers of e-manufacturing has become important for manufacturer in customer-oriented manufacturing to lure new customers along with retaining the old customers. It will also result in closing the gap between demand and supply of a product. This paper tries to assimilate the key enablers of e-manufacturing.

INTRODUCTION

Strictly speaking, the shop floor has always been viewed in isolation in a traditional manufacturing system. The factory operated as an autonomous unit, with its own set of goals and parameters. While this is a simple model for managing a factory, it is also inefficient and inflexible. Leaving the production department to its own devices means it is financially unaccountable at a time when stakeholders are demanding increased returns on net assets.

While traditionally manufacturing industry was product centric in nature, it is now slowly moving towards being customer centric. It has become imminently clear that making the customer the heart of the business process is essential. Relying on their products to increase profitability is no longer the answer. Customer focus is the buzzword in the manufacturing industry and the pivotal point in the business process. Manufacturers are fast accepting the fact that in order to boost sales and increase profits they need to pay more attention to customer satisfaction. Advanced Manufacturing Technology (AMT) was the next step towards being customer centric.

Though Advanced Manufacturing Technology (AMT) was thought as sufficient for customer driven market approach, success was only partial in today's internet-driven economy, that is, respond to customer demand in real-time. A real time response to the customer specific demand can only be met by integrating AMT with internet to form an integral part of e-business. This was the philosophy behind e-manufacturing.

E-manufacturing encompasses different meaning to different set of people. Just like "quality" has different definitions as perceived by different set of people, the same is the case with e-manufacturing. One of the popular definitions is by AMR Research [1]:

"The core of a manufacturing strategy is the technology roadmap for information transparency between the customer, manufacturing operations, and suppliers. An e-manufacturing strategy takes e-business processes, such as build to order and reliability-centered maintenance, and generates guidelines for implementing plant systems. The e-manufacturing strategy takes the e-business and manufacturing strategies and creates a roadmap for system development and implementation in the plant."

Another definition by Koç et al. [2]:

"E-manufacturing is a system methodology that enables the manufacturing operations to successfully integrate with the functional objectives of an enterprise through the use of Internet, tether-free (i.e., wireless, web, etc.) and predictive technologies."

E-manufacturing, may thus, be defined as the manufacturing environment wherein all inputs (men, material, machine, money, information) are processed for transforming into customer desired output through the deployment of distributed, flexible, open, reconfigurable, scalable, extendable communication, and data management systems.

The developments in manufacturing technology have been very well depicted in Figure 1.

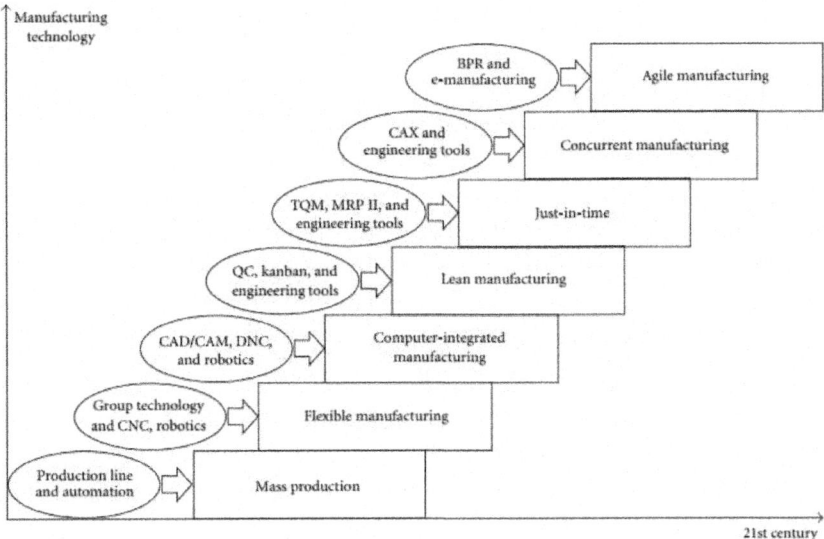

Figure 1: Development in manufacturing technology as adapted from Cheng and Bateman [3].

ENABLERS OF E-MANUFACTURING

Manufacturers can achieve the promise of e-manufacturing through the enablers that drive customer oriented build-to-order processes and nonstop operations.

Twelve key enablers have been identified for manufacturers to rapidly reap the benefits of e-manufacturing as discussed below.

Advanced Manufacturing Technology

Manufacturing has been defined by Lin and Nagalingam [4] and Wu [5]: "the organized process whereby products are made or created by various production activities from raw material".

Technology has been defined by Winner [6] from three dimensions: apparatus, referring to the equipment itself; technique, referring to the skills and knowledge necessary to use the equipment; and organization, referring to systems and structures of control and coordination.

Advanced word only reflects the phase of manufacturing. Thus, the word "Advanced Manufacturing Technology" came which only reflects the current phase of manufacturing technology.

Zammuto and O'Connor [7] defined AMT as a broad spectrum of computer-controlled automated process technologies. Beaumont et al. [8] described AMT more specifically as a group of computer-based technologies, including computer-aided design (CAD), computer numerical control (CNC) machines, direct numerical control (DNC) machines, robotics (RO), flexible manufacturing system (FMS), automated storage and retrieval system (AS/RS), automated material handling systems (AMHS), automated guided vehicles (AGV), bar coding (BC), rapid prototyping (RP), material requirement planning (MRP), statistical process control (SPC), manufacturing resource planning (MRP II), enterprise resource planning (ERP), activity-based costing (ABC), and office automation (OA).

As such Advanced Manufacturing Technology can be viewed as encompassing the computer and numerical-based apparatus (software and hardware) designed to accomplish or support manufacturing tasks. This definition excludes "managerial technologies" but may include, for instance, information networks for shop-floor data.

Web-Enabled Technologies and Services

E-manufacturing means giving the organization agility to react quickly to changes in the market, technology and customers by operating as a virtual enterprise. To exist as a virtual enterprise web-enabled technologies and services are required to integrate and synchronize information gathered based on inputs from customers, markets, and technology. Web-enabled decision-making portal shall perform dynamic optimization and synchronization based on the gathered information. Being one of the most important enablers of e-manufacturing, it should be based on a standard yet flexible protocol to assist manufacturing platforms.

Jin et al. [9] has discussed about web-enabled technologies and services in Networked Virtual Environments (Net-VEs) which provide an insightful, intuitive, and interactive system and allows effective communication among multiple users. The system allows engineers and designers to visualize, explore, manipulate, and interact with manufacturing applications in Net-VEs. Additionally, the industrial users can easily apply and share the manufacturing 3D data through web. By reducing costs and cycle time of product development, such an e-manufacturing system will speed up the major activities of manufacturing engineering including simulating manufacturing processes; optimizing assembly lines and workshops design; integrating labor and equipment, and hence, producing better quality products in a shorter time at more competitive price.

Stakeholder Interface and Feedback Mechanism

All stakeholders of a particular e-manufacturing environment should have an interface to interact with the system. The participation of stakeholders brings in the responsibility to ensure the correctness of input data and streamline the flow of information. The feedback mechanism thereby corrects any flaw in real-time by equating the current process with the correct process. Particularly, a Customer Relationship Management (CRM) package may be used for the purpose. In this age of the global market, with worldwide customers and remote offices, a web based software solution can help you manage your customers without any constraints of location or time.

As discussed by Lee et al. [10], an e-manufacturing strategy provides direct information exchanges between manufacturing and Customer Relationship Management (CRM) systems and supply chain management systems. It gives a complimentary set of process optimization methodologies for designing, operating, maintaining, or synchronizing the manufacturing operation using real-time information processing through stakeholders participation.

Quality Tools and Techniques

The efficient and consistent operation of plant throughout the enterprise is essential to attain high productivity which has always been a focus in manufacturing. As manufacturers consolidate through acquisitions and so have new facilities to operate worldwide, they must also learn how to make products with consistent quality and efficiency at each manufacturing location. This is where initiatives like lean manufacturing drive out waste, achieving nonstop operations for maximum efficiency and output, and where techniques like Six Sigma reduce variability in processes to ensure peak quality. Such quality tools and techniques need to be infused all along the e-manufacturing system to attain error-free working environment and thereby a quality product.

For achieving quality in e-manufacturing system, new standards need to be further identified and developed for the functional elements of the enabling tools for a manufacturing pant. These efforts should leverage the existing standard activities (i.e., IEEE 802.xx standard committees, MIMOSA, etc.) as discussed by Lee [11].

Rockwell Automation [12] published a white paper stating four competencies (design, operate, maintain, and synchronize) that are required for any manufacturer to be a world class manufacturing company.

Data Management System

Data management system basically includes mainframes which gather data from various machines and processes using internet and networking for seamless data flow. Since massive raw data is not useful unless it is reduced and transformed into information and knowledge for responsive actions, data mining tools for data reduction, representation and prediction adopted for plant floor data need to be developed. Corrective and proactive maintenance has to be implemented for the success of e-manufacturing.

As rightly depicted by Shivanand et al. [13], the large amounts of raw data collected during a manufacturing process are rendered useless, unless the data is gathered and transformed into some useful information which may be used to monitor a system.

Network Security

As the success of e-manufacturing depends on the seamless data flow across different agents of manufacturing, the security of network carrying this essential data becomes utmost critical for the working of e-manufacturing system. No compromise may be allowed on this front as any kind of interference with data may result in corrupt data being carried thereby resulting in wrong interpretation. The network has to be secured from any kind of virus or worm attack. The data which is vulnerable of being captured on internet while being transferred should be encrypted first and then transferred to the desired location.

Data/information security and vulnerability issues at the machine/product level need to be identified and corrected as pointed by Koç et al. [14].

Automation

Automation is the use of control systems and information technologies to reduce the need for human work in the production of goods and services. Automated manufacturing refers to the application of automation to produce things in the factory way. Most of the advantages of the automation technology have its influence in the manufacture processes.

The main advantage of the automated manufacturing are: higher consistency and quality, reducing the lead times, simplification of production, reducing handling, improving work flow and increasing the morale of workers when a good implementation of the automation is made. Modern, information-enabled automation architecture with associated integration management services helps in eliminating human-based errors.

The manufacturing enterprise is intensively deploying a host of hardware/ software automation/information technologies in order to face the changing societal environment pulled by the increasing customization of both goods and services as desired by customers and discussed by Morel et al. [15].

Pereira and Carro [16] discusses about advances in microelectronics and software allowing embedded systems to be composed of a large set of processing elements, and the trend is towards significant enhanced functionality, complexity, and scalability, since those systems are increasingly being connected by wired and wireless networks to create large scale distributed real-time embedded systems (DRES). Such embedded computing and information technologies have become at the same time an enabler for future manufacturing enterprises as well as a transformer of organizations and markets.

Maintenance

By using strategies that are predictive not just preventative, maintenance is seen as an opportunity to prevent unnecessary downtime, increase plant availability, and improve productivity. The efficient management of all company assets: materials, processes, and employees can be a part of globally networked monitoring systems to ensure nonstop operations and optimum asset productivity. Without such a solid, efficient foundation, it is not possible to achieve the twin goals of growth and profitability simultaneously. For the purpose a web-based software package may be used.

A CMMS software package maintains a computer database of information about an organization's maintenance operations, that is, CMMIS: computerized maintenance management information system. This information is intended to help maintenance workers do their jobs more effectively and to help management make informed decisions possibly leading to better allocation of resources. CMMS data may also be used to verify regulatory compliance.

E-Maintenance strategy should be adopted to integrate and synchronize the various maintenance and reliability applications to gather and deliver asset information where it is needed, when it is needed. Interconnectivity of the islands of maintenance and reliability information is embodied in E-Maintenance as mentioned by Shivanand et al. [13].

Supply Chain

Tools, e-fulfillment, and e-procurement solutions—direct transfer of customer order details to manufacturing personnel and equipment, configuration options, and close electronic connections to suppliers and partners to optimize the inbound supply chain in the face of changing demand. The integration of manufacturing operations into the greater supply chain, both upstream and downstream will facilitate proper synchronization. Competence in this area is best achieved after the other three are in place as discussed in Koç and Lee [17]. Only then is the plant truly ready to be fully coupled into an e-commerce-driven supply chain (see Figure 2).

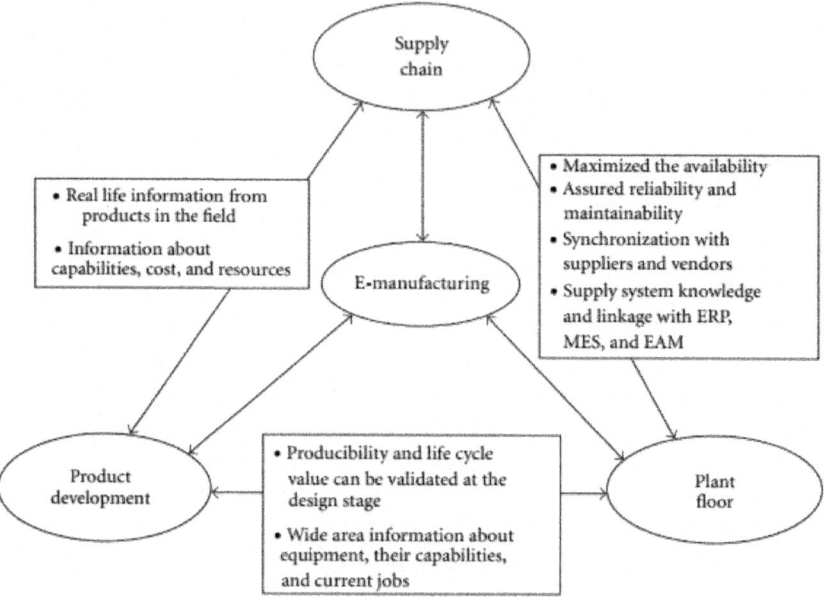

Figure 2: The transformation of e-Manufacturing for unmet needs in SCM as depicted in Koç and lee [17].

The importance of supply chain in an e-manufacturing system may also be gauged from the definition of e-manufacturing, "E-manufacturing is the vertical (business) and horizontal (supply-chain) integration of systems to ensure the correct dissemination of information throughout the value-chain of a business, making use of appropriate technology like the internet to ensure that real-time accurate information is available at all decision points throughout an organization and supply chain." As enumerated by Greeff et al. [18].

Web-based supply chain management (SCM) software may be integrated with e-manufacturing system for optimum productivity.

Interoperability of Software Systems

To achieve the goal of integrating information technology with manufacturing systems, the information must flow seamlessly from location to location without loss or corruption of data. The criticality further increases with increasing complexity of the type of data to be exchanged as more and more value-creating processes has to be inbuilt. As explained in NACFAM [19], the ability to achieve the desired levels of flexibility, efficiency, and responsiveness to exploit the full potential of integration is yet to be realized despite the advances in information and communications technologies. The software systems used by different companies need to follow a common standard of communicating with each other in such a way to eliminate the possibility of any type of information or data loss.

Synchronization

Seamless integration of all tools, techniques, and processes from shop floor to top floor is required to provide a basis for successful e-business. As discussed by Koç and Lee [17] synchronization of all integral tools, techniques and processes of e-business, e-manufacturing and e-maintenance needs to be addressed as shown in Figure 3.

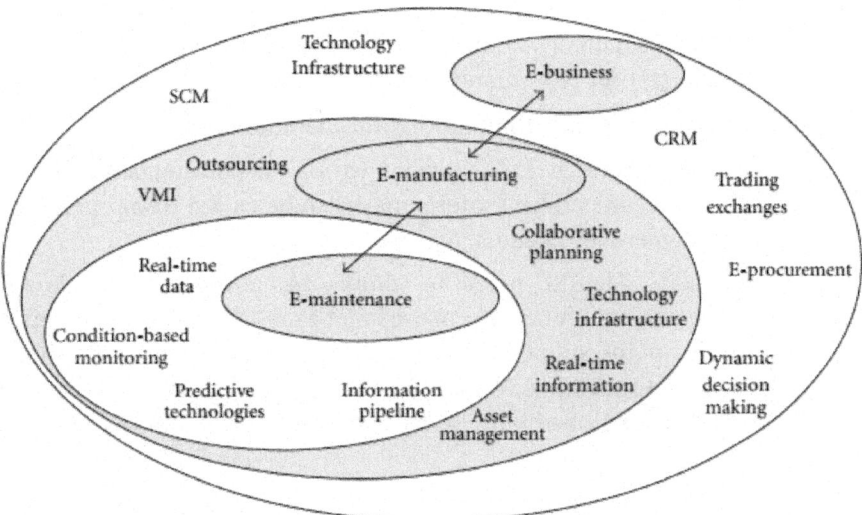

Figure 3: Integration of e-manufacturing with e-business and e-maintenance as deicted in Koç and Lee [17].

Educate and Train to Bring Awareness among Stakeholders

The success of e-manufacturing will ultimately depend on how well all stakeholders understand and become an integral part of it. For the purpose, it is necessitated to educate and train each stakeholder for all enabling tools and techniques. They must know their participative responsibilities to optimize the performance of e-manufacturing system.

e-Manufacturing is a transformation system that enables the manufacturing operations to achieve predictive near-zero-downtime performance as well as to synchronize with the business systems through the use of web-enabled and tether-free (i.e., wireless, web, etc.) infotronics technologies. It integrates information and decision-making among data flow (of machine/process level), information flow (of factory and supply system level), and cash flow (of business system level) as discussed in koç and lee [17].

CONCLUSION

The need for e-manufacturing may be summarized as methodologies and services for increasing overall asset utilization while minimizing unplanned maintenance and unexpected downtime.

E-manufacturing is more of philosophical nature than a concrete roadmap. The degree of focus on each enabler will vary according to a particular manufacturing environment. As such it becomes essential to prepare a guideline towards implementation of e-manufacturing.

Guidelines towards implementation of e-manufacturing:

- Constitute a team of experts from various departments and functions within the organization. Experts may also be called from outside of the company wherever needed.

- The experts should meet to draw the company's roadmap for e-manufacturing. It must be ensured that each section of the organization has a vision for success based on enablers of e-manufacturing. The common and specialized benefits must be clearly outlined. A framework needs to be developed with all stakeholders' action outlined in a timely manner.

- The internet should be an inherent part of the manufacturing system, and all human resource of the company must adopt it.

- Engineers along with first line workers being an integral part of the shop-floor decision-making process, their experience and vision shall be invaluable in seamless integration of shop floor with e-manufacturing.

- The head of the company should also be the head of e-manufacturing system. Any strategy towards benefit of the company must have e-manufacturing as a key element of that broader strategy.

- All stages of the roadmap should be measured towards its success and failure. Measurable targets should be placed to determine the real savings and efficiencies from a transparent company.

- Existing foundations of the company should be used to build upon for incremental successes along the way. An e-manufacturing strategy will help the organization embrace information transparency, fostering operational excellence while cutting down on wastes.

- Evaluate and streamline the existing processes with customers, suppliers, distributors, and others to optimize their effectiveness in e-manufacturing system.

- Educate and train all stakeholders in pursuing goal of increasing efficiency on every front and harnessing all possible benefits from e-manufacturing.

- Organizations are now trying to create a wonderful experience for their customers and are focusing more on the experience that their customer has which stems from a deep desire on the customer's part to have a good relationship with the company. This will ensure a better managed e-manufacturing system within e-business.

REFERENCES

1. "An e-manufacturing strategy needs to be developed from the manufacturing strategy," AMR Research, Inc, August 2000.

2. M. Koç, J. Ni, and J. Lee, "Introduction of e-manufacturing," in Proceeding of the International Conference on Frontiers on Design and Manufacturing, pp. 43–47, Dalian, China, July 2002.

3. K. Cheng and R. J. Bateman, "E-manufacturing: characteristics, applications and potentials," Progress in Natural Science, vol. 18, pp. 1323–1328, 2008.

4. G. Lin and S. Nagalingam, CIM Justification and Optimisation, Taylor and Francis, London, UK, 1st edition, 2000.

5. B. Wu, Manufacturing Systems Design and Analysis, Chapman & Hall, London, UK, 2nd edition, 1994.

6. L. Winner, Autonomous Technology, MIT Press, Cambridge, Mass, USA, 1977.

7. R. Zammuto and E. O'Connor, "Gaining advanced manufacturing technologies' benefits: the roles of organization design and culture," Academy of Management Review, vol. 17, pp. 701–728, 1992.

8.   N. Beaumont, R. Schroder, and A. Sohal, "Do foreign-owned firms manage advanced manufacturing technology better," International Journal of Operations & Production Management, vol. 22, no. 7, pp. 759–771, 2002.

9.   L. Jin, I. A. Oraifige, P. M. Lister, et al., "E-manufacturing in networked virtual environments," inProceedings of the IEEE International Conference on Systems, Man and Cybernetics, vol. 3, pp. 1845–1849, October 2001.

10.   J. Lee, A. Ali, and M. Koç, "E-manufacturing-its elements and impact," in Proceedings of the Annual Institute of Industrial Engineering Conference, vol. 23, Dallas, Tex, USA, May 2001.

11.   J. Lee, "E-manufacturing—fundamental, tools, and transformation," Robotics and Computer-Integrated Manufacturing, vol. 19, no. 6, pp. 501–507, 2003.

12.   "Rockwell Automation E-manufacturing Industry Road Map,," http://www.rockwellautomation.com/.

13.   H. K. Shivanand, N. V. Nanjundaradhya, and P. Kammar, "E-manufacturing a technology review," inProceedings of the World Congress on Engineering, vol. 2, London, UK, July 2008.

14.   M. Koç, J. Ni, J. Lee, et al., "Introduction of e-manufacturing," in Proceedings of the International Conference on Frontiers on Design and Manufacturing, pp. 1–10, July 2005.

15.   G. Morel, P. Valckenaers, J. M. Faure, et al., "Manufacturing plant control: challenges and open issues," in Proceedings of the 16th IFAC Triennial World Congress, 2005.

16.   C. E. Pereira and L. Carro, "Distributed real-time embedded systems: recent advances, future trends and their impact on manufacturing plant control," Annual Reviews in Control, vol. 31, no. 1, pp. 81–92, 2007.

17.   M. Koç and J. Lee, "E-manufacturing-fundamentals, requirements and expected impacts," International Journal of Advanced Manufacturing Systems, vol. 6, no. 1, pp. 29–46, 2003.

18.   G. Greeff, R. Ghoshal, and S. Mackay, Practical E-Manufacturing and Supply Chain Management, chapter 1, Elsevier, 2004.

19.   NACFAM, Exploiting E-Manufacturing: Interoperability of Software Systems Used By U.S. Manufacturers, National Coalition for Advanced Manufacturing, Washington, DC, USA, 2001.

Chapter 9

BIOCOMPATIBILITY OF ADVANCED MANUFACTURED TITANIUM IMPLANTS— A REVIEW

Alfred T. Sidambe

Bioengineering & Health Technologies Group, School of Clinical Dentistry, University of Sheffield, 19 Claremont Crescent, Sheffield S10 2TA, UK;

ABSTRACT

Titanium (Ti) and its alloys may be processed via advanced powder manufacturing routes such as additive layer manufacturing (or 3D printing) or metal injection moulding. This field is receiving increased attention from various manufacturing sectors including the medical devices sector. It is possible that advanced manufacturing techniques could replace the machining or casting of metal alloys in the manufacture of devices because of associated advantages that include design flexibility, reduced processing costs, reduced waste, and the opportunity to more easily manufacture complex or custom-shaped implants. The emerging advanced manufacturing approaches of metal injection moulding and additive layer manufacturing are receiving particular attention from the implant fabrication industry because they could overcome some of the difficulties associated with traditional implant fabrication techniques such as titanium casting. Using advanced manufacturing, it is also possible to produce more complex porous structures with improved mechanical performance, potentially matching the modulus of elasticity of local bone. While the economic and engineering potential of advanced manufacturing for the manufacture of musculo-skeletal implants is therefore clear, the impact on the biocompatibility of the materials has been less investigated. In this review, the capabilities of advanced powder manufacturing routes in producing components that are suitable for biomedical implant applications are assessed with emphasis placed on surface finishes and porous structures. Given that biocompatibility and host bone response are critical determinants of clinical performance, published studies of *in vitro* and *in vivo* research have been considered carefully. The review concludes with a future outlook on advanced Ti production for biomedical implants using powder metallurgy.

INTRODUCTION

Materials used for biomedical applications exist in different forms and they must possess specific properties to fulfil this role [1]. Materials such as metals are commonly used as implants and such metals have to possess properties which will enable them to function inside the human or animal body [2]. Biomaterials, as they are also known, are expected to have biomechanical properties which are comparable to those of autogenous tissues without side effects. The properties which determine whether a material is suitable for biomedical implant applications include biocompatibility, bioadhesion, biofunctionality and corrosion resistance [2]. In order to ensure safety and to have the desired results, implants and other devices intended for biomedical use are regulated by different bodies globally such as the U.S. Food and Drug Administration (FDA) and the International Standards Organisation (ISO) [3]. The main metallic biomaterials are stainless steels, cobalt alloys, titanium and titanium alloys [1]. This review has been limited to the study of titanium and its alloys because this is a metal whose widespread use has been limited by its high cost due to the multi-step Kroll extraction process of the Ti raw material [4]. Advanced powder manufacturing routes such as metal injection moulding (MIM) have emerged as techniques that can minimise the cost of titanium implant production.

TITANIUM AS IMPLANT MATERIAL

Titanium and titanium alloys exhibit a high specific strength [5], which makes titanium an excellent choice for biomedical applications [6]. Furthermore, titanium is considered to be biocompatible because it has a low electrical conductivity which contributes to the electrochemical oxidation of titanium leading to the formation of a thin passive oxide layer [7]. The oxide layer in turn leads to a high resistance to corrosion. This protective passive layer is retained at pH values of the human body [8] due to titanium having an oxide isoelectric point of 5–6 [1]. In aqueous environments Ti and its oxides have low ion-formation tendency and low reactivity with macromolecules [9]. Titanium alloys are used in biomedical implant devices which replace damaged hard tissue. Some examples of Ti uses in biomedical applications are dental and orthopaedic implants, artificial hearts, pacemakers, artificial knee joints, bone plates, cardiac valve prostheses, screws for fracture fixation, artificial hip joints [1] and cornea backplates [10]. Titanium and titanium alloys have therefore been used widely as biomedical implant materials since the early 1970s and the implants have been available as machined and cast components. The alloys that are preferred for the fabrication of titanium implants are commercially pure titanium (CP-Ti) and titanium alloy Ti6Al4V (Ti-64). CP-Ti has a higher resistance to corrosion and is widely regarded as the most biocompatible metal because of a stable and an inert oxide layer which spontaneously forms when its surface is exposed to oxidising media [1]. Almost all commercially available permucosal dental implants are made from CP-Ti as a result of the pioneering research of Brånemark and his co-workers [11].

The CP-Ti and Ti-64 manufactured via the traditional routes (such as strips, sheets, plates, bars, billets, forgings and wires) are specified according to the American Society for Testing and Materials (ASTM) as grades 1 to 5. Grades 1 to 4 are the unalloyed CP-Ti and grade 5 is the alloyed Ti-64. Table 1 summaries the mechanical properties of titanium according to the ASTM standards F67 [12] and F136 [13] for bars, billets and forgings. Grade 2 titanium is the main unalloyed Ti used in dental implant applications. Grade 2 Ti has a minimum yield strength of 275 MPa and this is the equivalent of yield strength in heat-treated austenitic stainless steels. Grade 5 Ti-64 is the most widely used titanium alloy in biomedical implants where high strength is required [1]. As it can be seen form Table 1, CP-Ti has lower strength whilst Ti-64 is an $\alpha + \beta$ alloy which offers a higher strength [14].

Table 1. Selected mechanical requirements properties of titanium bar for implant [1]

Material	Specification	Tensile strength (MPa)	0.2% Proof stress (MPa)	Elongation (%)	Elastic modulus (GPa)
CP-Ti	ASTM F67 Grade 1	240	170	24	103–107
-	ASTM F67 Grade 2	345	275	20	103–107
-	ASTM F67 Grade 3	450	380	18	103–107
-	ASTM F67 Grade 4	550	483	15	103–107
Ti6Al4V	ASTM F136 Grade 5	860	795	10	114–120

MPa = megapascal, GPa = gigapascal.

As much as CP-Ti and Ti-64 have a number of advantages which have already been mentioned above, they do however have some disadvantages as far as implant applications are concerned in that they both have low wear resistance properties. There have also been concerns that the vanadium contained in the Ti-64 alloy is cytotoxic [2] which means that this alloy is limited to certain appropriate applications and devices [13]. CP-Ti and Ti-64 have an elastic modulus which is considered to be high when compared to the bone [2]. These factors therefore limit the application of CP-Ti and Ti-

64 because of the mismatch of the elastic (or Young's) modulus between the implant and the bones [15]. A low elastic modulus is desirable in implants because it helps to avoid stress shielding and the associated bone resorption [16]. As a result, new titanium alloy compositions which are specifically tailored for biomedical applications have been developed. The first generation of these biomedical implant alloys have included Ti-6Al-7Nb and Ti-5Al-2.5Fe, two alloys with properties similar to Ti-64 that were developed to address concerns relating vanadium's cytotoxicity [2]. Titanium alloys containing Nb, Ta and Zr (β stabilising elements) with a low modulus, high strength, resistance to corrosion and biocompatibility have also been developed for biomedical implant applications.

Another solution to the problems associated with the high titanium elastic modulus has been to use advanced manufacturing processes such as additive layer manufacturing (ALM) to make highly porous titanium structures of CP-Ti and Ti-64. These porous structures can be tailored to have excellent mechanical properties similar to those of human bone and are usually designed to facilitate bone ingrowth [17]. The elastic modulus of titanium is about 114 GPa while that of cancellous and cortical bone range from 0.5 GPa to a maximum of 20 GPa [18]. The mechanical properties and behaviour of titanium implants can be varied via the volume fraction and size distribution of the pore structures [19] and it has been shown that the elastic modulus of porous titanium decreases with increasing pore size under a compressive force [18].

The increased interest in the advanced manufacturing of titanium has led to the subsequent need for regulations and standards and in response the ASTM and the International Organization for Standardization (ISO) have recently published new standards for ALM and MIM titanium. Table 2 shows the chemical composition requirements for MIM and ALM titanium according to ASTM standards F2989, F2885 [20], F2924 [21], F3001 [22] and ISO 22068 [23] standards. Titanium ASTM F2989 "MIM 1" has the highest purity with 0.20 wt % maximum iron content, which is lower than "MIM 2" and "MIM 3" (0.30 wt % maximum). In comparison to the chemical properties of ASTM F67, the chemical requirements are unchanged except for the oxygen in Grade MIM 3 which is set at 0.3 wt % maximum whereas it is 0.35 wt % in ASTM F67. When comparing the alloyed Ti-64 ASTM standards, the maximum oxygen level is set at 0.13 wt % extra-low interstitial (ELI) for the bars, billets and forgings (ASTM F136). In MIM the maximum oxygen level is 0.2 wt % because in processing MIM there is always oxygen pickup during sintering.

As it can be seen from Table 2 there are currently two alloyed Ti grades made by ALM that are in the standards and that can be used for surgical implant applications. The ASTM F2924-14 [21] and F3001-14 [22] standards cover additively manufactured Ti6Al4V components using full-melt powder bed fusion such as electron beam melting and laser melting [22]. Table 3 shows the minimum mechanical properties of Ti manufactured by ALM and MIM according to ASTM F2924-14, ASTM F3001-14, ASTM F2989-11, F2885-11 and ISO 22068.

Table 2: Chemical composition requirements for metal injection moulding (MIM) and additive layer manufacturing (ALM) titanium according to American Society for Testing and Materials (ASTM) standards

Specification	Chemical composition (wt %)							
	Al	V	Fe	C	N	O	H	Y
ASTM F2989 MIM 1	-	-	<0.2	<0.08	<0.03	<0.18	<0.015	-
ASTM F2989 MIM 2	-	-	<0.3	<0.08	<0.03	<0.25	<0.015	-
ASTM F2989 MIM 3	-	-	<0.3	<0.08	<0.05	<0.30	<0.015	-
ISO 22068 MIM Ti-400	-	-	-	<0.2	<0.1	<0.4	-	-
ASTM F2885 Grade 5	5.5–6.75	3.5–4.5	<0.30	<0.08	<0.05	<0.20	<0.015	<0.005
ISO 22068 MIM-Ti6Al4V-600	5.0–7.0	3.0–5.0	-	<0.2	<0.1	<0.4	-	-
ASTM F2924 ALM Ti6Al4V	5.5–6.5	3.5–4.5	<0.3	<0.08	<0.05	<0.20	<0.015	-
ASTM F3001 ALM Ti6Al4V (ELI)	5.5–6.5	3.5–4.5	<0.25	<0.08	<0.05	<0.13	<0.012	<0.005

In comparison to requirements for unalloyed titanium bars, billets and forgings, the tensile strength and yield strength requirements are higher in ASTM F2989 MIM whereas the elongation requirements are lower. This is because in the as-sintered state Ti MIM carries some residual porosity [24]. The tensile strength and yield strength requirements in Ti-64 MIM are lower than in bars, billets and forgings whereas it is the same in Ti-64 ALM. The requirements for elongation are lower with Ti-64 ALM.

Table 3: Mechanical properties of CP-Ti MIM, Ti-64 MIM and ALM

Material	Specification	Tensile strength (MPa)	0.2% Proof stress (MPa)	Elongation (%)	R of A * (%)
MIM CP-Ti	ASTM F2989 MIM 1	370	315	23	25
-	ASTM F2989 MIM 2	420	360	17	20
-	ASTM F2989 MIM 3	495	390	10	15
-	ISO 22068 MIM Ti-400	500	400	5	
MIM Ti6Al4V	ASTM F2885 Grade 5	780	680	10	15
-	ISO 22068 MIM-Ti6Al4V-600	800	600	3	-
ALM Ti6Al4V	ASTM F2924 ALM Ti6Al4V	895	825	10	15
ALM Ti6Al4V	ASTM F3001 ALM Ti6Al4V (ELI)	860	795	8	25

* R of A = Reduction of Area.

ADVANCED POWDER PROCESSING OF TITANIUM

As it was mentioned earlier, titanium production is hampered by the high cost in traditional manufacturing processes, and poor workability for complex shape production. This has led to numerous investigations of various potentially lower-cost processes [4]. Although there are various powder metallurgy routes of processing titanium as have been listed by Froes [25], this review is confined to the relatively new areas of ALM and titanium MIM.

Advanced manufacturing routes such as ALM and Ti MIM are processes which offer design flexibility and cost-saving respectively, for fabricating products that have complicated shapes with a very high accuracy of size. ALM and MIM are distinct from traditional machining techniques, which mostly rely on the removal of material by methods such as cutting or drilling (subtractive manufacturing) [26]. Thus, ALM and MIM are increasingly being employed as processes for fabricating surgical implants including orthopaedic and dental implants. With increasing demand for medical implantation, there is increasing need for implants that offer reliability, adequate mechanical properties and that offer unique properties (such as patient specific implants) as well as comfort. MIM is a processing route that offers reduction in costs, with the added advantage of near net-shape fabrication as reported by Ferri *et al.*[27] and Ebel *et al.* [28]. On the other hand ALM is used to make patient specific, complex, cellular and functional mesh arrays implants or bone substitutes [29].

The aim of this review is to summarize existing literature and report on the use of advanced manufacturing to fabricate titanium alloy implants. Emphasis is placed on the surface finishes, porous structures and various *in vitro*and *in vivo* biocompatibility studies that have been carried out on the advanced manufactured implants. The development of new processes and materials does not typically include biocompatibility testing until the prototype stage is reached. Regulation agencies such as the Food and Drug Administration (FDA) in the USA require biocompatibility testing per ISO 10993 or ASTM F748 prior to device approval [30]. Therefore there is a need to carry out biocompatibility testing in any material or new material processing method [30].

Additive Layer Manufacturing

Additive layer manufacturing encompasses a group of technologies that have emerged in the last decade. In ALM, the material is added one cross-sectional layer at a time to create an object [31]. ALM uses the additive method to make a three-dimensional solid object of almost any shape from a computer-aided design (CAD) model. The CAD file model which constitutes the part geometry is created and once optimised, the CAD file is then "sliced numerically" into the layer thickness the machine will build in. After that it is transferred to the ALM machine's software and loaded to the ALM machine allowing a file based build to begin. Layers of material are laid down successively with each layer corresponding to a different shape. Figure 1 illustrates the flow diagram of ALM system from CAD file through to component manufacture.

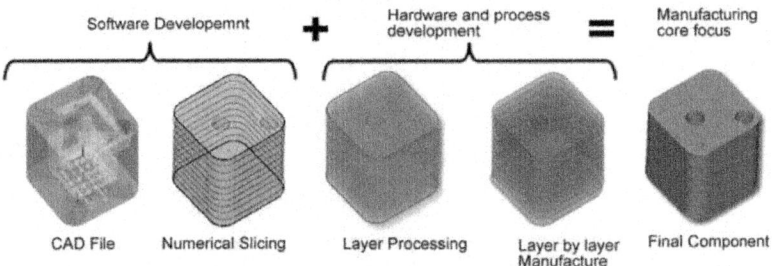

Figure 1: Flow diagram of ALM system from computer-aided design (CAD) file through to component manufacture.

ALM technologies include fused deposition modelling, laser micro-sintering, direct metal laser sintering (DMLS), three-dimensional (3-D) laser cladding, electron beam melting (EBM), and electron beam sintering (EBS) [26]. This review looks in detail at the two systems which have the capability to manufacture Ti, namely the EBM and DMLS and briefly at other ALM techniques.

Electron Beam Melting (EBM)

Electron beam melting (EBM) is one of the additive manufacturing techniques mainly used for metallic biomaterials. The EBM system manufactures parts by melting the metal powder layer by layer using a magnetically directed electron beam (up to 3 kW) under a high vacuum atmosphere [32]. It is for this reason that EBM is particularly suited for the manufacturing of titanium and titanium alloy implants. Figure 2 is a schematic drawing of an electron beam melting system.

Part of the challenges of using EBM in the manufacturing of surgical implants is to optimise the surface finish of the final components. This highly depends on the EBM processing parameters such as beam current, part orientation and powder particle size. It is widely known and accepted that the surface topography of biomedical implants affects the biocompatibility because it influences the cell attachment, proliferation, and differentiation. The performance of biomedical implants and their biocompatibility depends very much on the initial interaction between surfaces of the implants and the biological environment [33]. As a result, there are a number of studies that have been carried out to determine the biocompatibility of titanium implants manufactured via EBM.

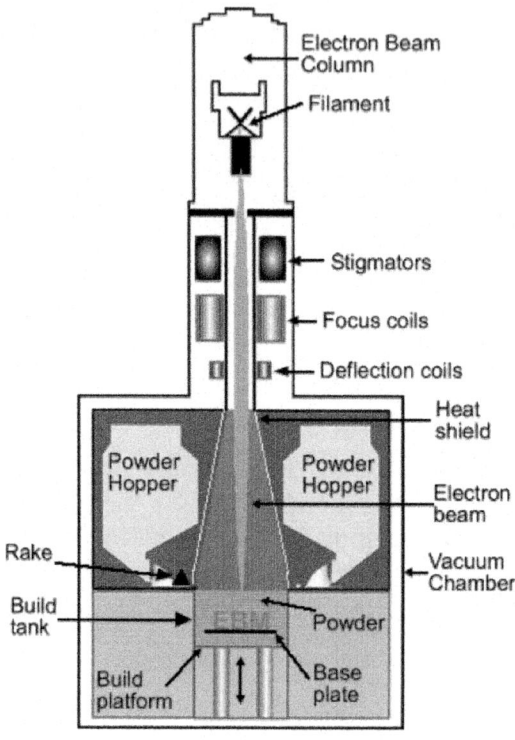

Figure 2: Schematic drawing of an electron beam melting system.

Ponader *et al.* [33] manufactured Ti-64 components with different surfaces and porosity using EBM and then they assessed, *in vitro*, the surface matrices for cell attachment, proliferation, and differentiation using human fetal osteoblasts (hFOB 1.19). In that study the cell proliferation was found to be more pronounced in the Ti-64 samples which were compact than on the porous samples. These Ti-64 compacts consisted of adherent partly molten titanium particles on the surface, leading to a surface roughness (Ra) of ≤24.9 µm. In porous samples with highly rough surfaces (Ra ≥ 56.9 µm) there was reduced proliferation of the hFOB cells. The authors found that cell differentiation markers were not influenced by the surface roughness. Surface characteristics of titanium could easily be changed by EBM in order to further improve proliferation [33]. This study also concluded that the wear resistance must be optimized and that further investigations were necessary to clarify whether the observed *in vitro* biocompatibility can also be observed *in vivo*, where bone tissue can grow into the EBM structures. The authors then carried out a follow on study where they assessed EBM smooth compacts and porous Ti-64 structures as scaffolds for bone formation, *in vivo* [17]. The implants were placed into defects in the frontal skull of 15 domestic pigs. After surgery, X-Ray photographs were taken to analyse the microscopic structure of the direct contact between bone and implant surfaces in intervals of 14, 30, and 60 days. The results showed that bone ingrowth was increasing with the intervals as follows: around 14% after 14 days, 30% after 30 days and 46% after 60 days in both smooth compact and rough porous implants. There was less bone-implant contact around compact specimens (6%) than around porous specimens (9%) after the 60 days. In the study the authors succeeded in demonstrating the suitability of highly porous titanium implants as scaffolds for bone ingrowth especially in orthopaedic and maxillofacial applications [17].

Whilst Ponader *et al.* [17] *in vivo* study showed relatively low osseointegration within the first 60 days, another *in vivo* study which examined the bone-to-implant contact of EBM Ti-64 implants was carried out by Thomsen *et al.* [34] and this showed the bone-to-implant contact as 29%–41% after 6 weeks in rabbits. After producing EBM Ti-64 implants with an increased surface roughness, the authors machined a selected number of samples and also included conventionally produced bulk Ti-64 material in their study. The results of the early bone response in rabbits showed that the as-EBM Ti-64 implants showed no significant difference in tissue response when compared to the conventional wrought titanium alloy implants and the machined EBM Ti-64. The as-EBM Ti-64 implant specimens had an area surface roughness (S_a) greater than 15 μm. Also revealed in this study were the results of the chemical and mechanical properties of the electron beam-melted Ti-64 material which were within the ASTM F136 specifications. This study was done prior to the issuing of the ASTM 3001 standard.

The biocompatibility of polished, unpolished and porous EBM Ti-64 was also studied by Haslauer *et al.* [29] using *in vitro* cell culture. The authors compared the cellular response in discs which were seeded with 20,000 human adipose-derived adult stem cells (hASCs). The assessments showed that the hASCs were alive on all discs after 8 days, that cellular proliferation on porous EBM Ti-64 discs was increased compared to solid polished and unpolished EBM discs and that the release of the pro-inflammatory cytokines (IL-6 and IL-8) was lower for porous EBM discs than for other discs. Additionally, IL-6 and IL-8 releases at day 7 were lower for porous EBM discs than for other discs. The authors were satisfied that the biocompatibility of the EBM Ti-64 discs was comparable to that of the commercial implant types.

In a study which involved the surface modification of implants via coating, Li *et al.* [35] fabricated porous Ti-64 implants via the EBM process after which they used the biomimetic approach to coat the surfaces. The *in vitro* and *in vivo* biocompatibilities were evaluated for implants with and without biomimetic apatite coating. The results of the *in vitro* biocompatibility of the EBM porous titanium were positive for cell attachment, proliferation and cell morphology. The comparison of implants with and without biomimetic coating showed that cell proliferation in porous EBM Ti-64 implants could match that of coated implants. Similarly, the *in vivo* histological analysis showed a comparable rate of both ingrowth and bone formation between EBM Ti-64 and coated Ti-64 after 12 weeks. The analysis of the mechanical properties showed that the EBM porous Ti-64 had a Young's modulus that is similar to cortical bone (14.5 to 38.5 GPa).

In a study that included another long-term *in vivo* experiment, Palmquist *et al.* [36] evaluated the osseointegration of porous EBM Ti-64 and solid EBM machined Ti-64 cylindrical and disk-shaped implants. The Ti-64 was implanted in sheep, bilaterally in the femur and subcutaneously in the dorsum. After retrieval 26 weeks later, the results showed that both porous and solid implants were osseointegrated and that high bone–implant contact of 57% was observed throughout the porous implant. However, in discussing the effects of surface topography, the authors suggested that it is not always certain that specific biological effects can be assigned to specific surface property.

The above assertion by Palmquist *et al.* [36] is confirmed by the fact that, although they are interlinked, results of the biocompatibility EBM of titanium alloys listed above are varied. Cell proliferation has been shown to be more pronounced in specimens which are compact and with a lower surface roughness, whereas in another case the authors claim the EBM Ti-64 had the biocompatibility properties which matched reference samples [35] and in another case the authors found more proliferation in rougher specimens compared to the reference samples [29]. Nevertheless, the literature survey has revealed to a certain extent that EBM of titanium can be adopted for the fabrication of custom orthopaedic implants. The process can be used in designing implants with specific surface roughnesses and porosity. The biocompatibility studies have shown how the design flexibility can be correlated with bone ingrowth and cell proliferation, for example. The results open up the possibilities of wider use of EBM titanium to reconstruct specific bone defects.

Direct Metal Laser Sintering (DMLS), Selective Laser Melting/Sintering

Direct metal laser sintering (DMLS) is an additive metal fabrication technology that was developed by EOS who are based in Munich, Germany. DMLS is often also referred to using the terms selective laser sintering (SLS) or selective laser melting (SLM).

In the DMLS system, a high-powered optic laser with a power of 200 W to 400 W is used to fuse metal powder into a solid component based on a 3D CAD file. In a similar manner to the EBM system, components are built from layer to the next layer using the additive method with the layer thickness typically being 20 μm. Figure 3 shows the schematic overview of SLM form of DMLS cycle.

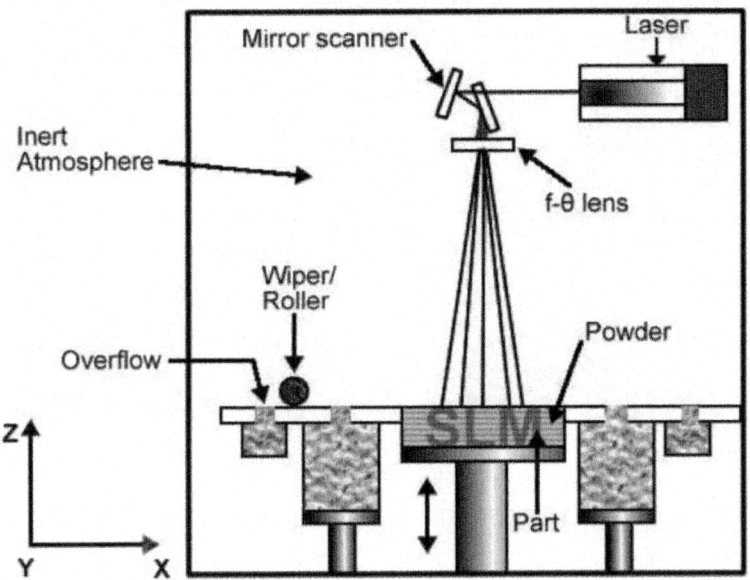

Figure 3: Schematic overview of selective laser melting (SLM) cycle.

The DMLS therefore carries the challenges of manufacturing of biomedical implants with the desired surface finish without the need to post process by polishing *etc.* as on the EBM system. One of the studies that have been carried out in order to investigate the suitability of the titanium DMLS for biomedical implants fabrication was presented by Hollander *et al.* [37]. They presented results which demonstrated the effect of different surface properties on Ti-64 "bone substitutes" made via SLM. In particular, the biocompatibility of

SLM Ti-64 material was studied using the primary human osteoblasts cells (HOB). Comparisons were made of the SLM surfaces with the commercial Thermanox® (Nalge Nunc Int., New York, NY, USA) control and conventional bulk titanium. The results showed that the cultured cells attached and proliferated on SLM substrates and the activity of alkaline phosphatase (AP) was also shown to increase after 7 days, but decrease after 14 days [37]. The authors concluded that the increased metabolic activity of osteoblasts on SLM discs compared to the controls may have been due to the greater surface area on the SLM material, which took longer to be covered by the cells. Figure 4 shows the example of an exact computerised tomography (CT) based complex human vertebra processed by the authors using SLM.

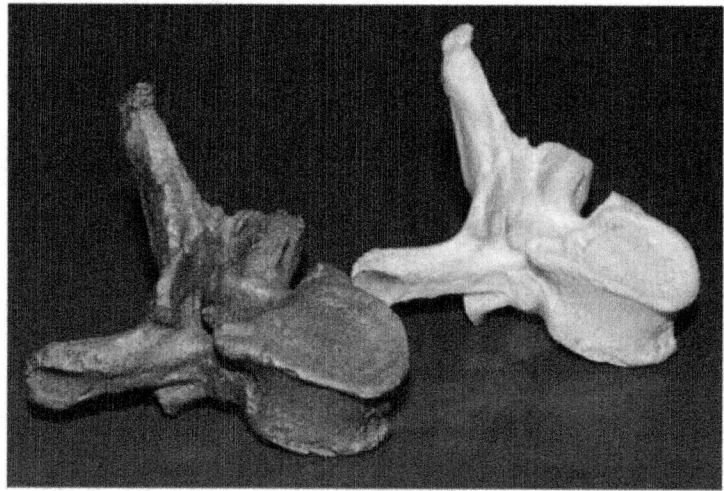

Figure 4: Example of an exact CT-based part of a complex human vertebra processed by SLM using additive manufacturing [37].

Using a DMLS variant called the Selective Laser Sintering (SLS) method, Hollister *et al.* [38] manufactured craniofacial and temporomandibular joint (TMJ) scaffolds for craniofacial reconstruction. The authors developed two interconnected porous architecture designs, one being interconnected spherical pores and the other being interconnected cylindrical pores whose porosity was ranging from 50% to 70%. They used various biomaterials which included an unspecified grade of titanium and presented in the results initial *in vitro* and *in vivo* test data showing substantial bone ingrowth (between 40% and 50% at 6 weeks, between 70% and 80% at 18 weeks) for all scaffolds. The results of the *in vivo* tests showed that the scaffolds, including titanium, were able to support bone regeneration via delivery of BMP-7 transduced human gingival fibroblasts in a mouse model. Furthermore, the titanium scaffolds had mechanical property values lying in between those of cortical and trabecular bone and also supported bone and cartilage regeneration needed to reconstruct craniofacial structures.

Therefore in implantology involving metallic biomaterials like Ti it is important to understand the role of specific surface properties when bone is in contact with an implant. It is for this reason that studies have also been carried out to understand early human bone response to Ti implants, with the more recent studies looking at the advanced manufacturing routes such as DMLS. The early human bone response to a DMLS Ti-64 implant surface has been reported by Mangano *et al.* [39] after they carried out a study where micro-implant was inserted into the anterior mandible of a patient during conventional implant surgery of the jaw as part of the *in vivo* study. After two months of unloaded healing in this study, the micro implant and surrounding tissues were extracted and the structure of tissue was studied by microscopy. The histology showed the peri-implant bone in close contact with the surface of the implant. The mean of bone-to-implant contact was 69.51%. The authors concluded that DMLS Ti-64 is a promising alternative to conventional implant surface topographies [39].

Whilst using *in vitro* studies to demonstrate osseointegration of DMLS titanium, Warnke *et al.* [40] cultured human osteoblasts on SLM-produced Ti-64 mesh scaffolds. Evaluation of the cell occlusion of pores with different widths (0.45 to 1.2 mm) was carried out. The biocompatibility results showed that the osteoblasts had a well-spread morphology and also had multiple contact points. It was found that pore overgrowth increased after 6 weeks of culture where the pore widths were 0.45 and 0.5 mm, and in the course of 3 weeks where the pore widths were 0.55, 0.6, and 0.7 mm. No pore closure was observed on pores of width 0.9–1.2 mm. The authors also found that porosity and maximum compressive load at failure increased and decreased with increasing pore width, respectively. In summary, their scaffolds were biocompatible, and pore width influenced pore overgrowth and resistance to compressive force [40].

The literature survey has revealed to a certain extent that DMLS of titanium can be adopted for the fabrication of custom orthopaedic implants. The results of *in vivo* tests showed that the DMLS scaffolds with unpolished and unmodified surfaces were able to support bone regeneration and that DMLS of titanium as an advanced manufacturing technology is a promising alternative to conventional implant surface topographies. On the other hand the survey has revealed that porous Ti DMLS scaffolds are biocompatible, and that pore width can influence growth around the pores and the resistance to compressive force.

Metal Injection Moulding

Metal injection moulding is a processing route that offers reduction in costs for Ti biomaterials and implants, with the added advantage of near net-shape components. The small to medium sized components with complex geometries that can be manufactured via MIM have the potential to be fully exploited by the implant industry.

The MIM cycle starts with preparation of a feedstock which consists of mixing fine metallic powder and a binder material. The resulting feedstock is then granulated and an injection moulding machine is used to inject the feedstock into a mould cavity under an elevated temperature (<200 °C) and pressure. Once in the mould the molten feedstock cools and solidifies to produce a green part. The binder is then removed and what is left behind is a highly porous brown part. The brown part is then sintered and shrinks, typically to >95% of the pore-free density (PFD) [24]. More details about MIM have been published elsewhere [41] and the flow diagram for the MIM process is shown in Figure 5.

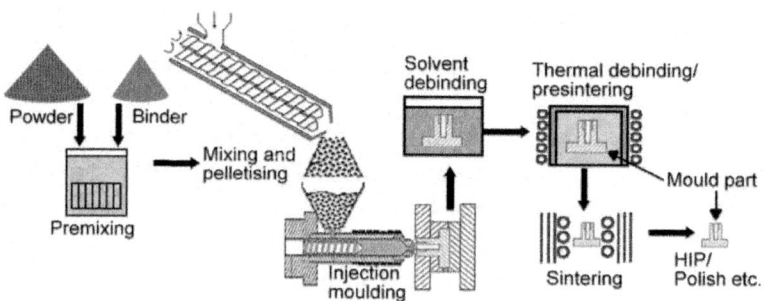

Figure 5: Metal injection moulding flow diagram.

MIM in manufacturing is well established but the processing of titanium and titanium alloys in this field is still in its infancy because the process has had to depend on high quality starting powders, carbon free binders and improved furnaces for sintering in order to reduce contamination of the titanium. With the necessary improvements, the interest in the manufacturing of titanium for biomedical applications has gathered pace in recent years. Among the challenges in titanium MIM is the production of components with desirable surface finishes, mechanical and chemical properties. In MIM, the quality of the surface finish highly depends on the processing parameters such as mould tool finish and starting powder particle size distribution.

Among the studies that have shown the biocompatibility of implantable MIM Ti is a publication by Sago *et al.* [42]. In the study the authors manufactured Ti-64 devices via MIM and presented the results of the mouldability, the microstructure of the alloy, the mechanical properties as well as the biocompatibility. Their results showed that MIM can produce Ti-64 alloys which meet the specification requirements of implantable ASTM F1472 in the hot isostatically pressed (HIPped) condition. The samples were shown to have passing results for a series of biocompatibility tests, comparable to the wrought grade. Biocompatibility testing for MIM Ti-64 was carried out for hemocompatibility (ISO 10993-4), cytotoxity (ISO 10993-5), sensitization (ISO-10993-10), irritation (ISO 10993-10), systemic toxicity (ISO 10992-11), and implantation (ISO 10993-6) [42].

In a later study, Sago et al. [43] also found that the chemical, metallurgical, and mechanical properties of the MIM Ti-64 met the recently published property specifications ASTM F2885 for MIM implantable grades. The samples they manufactured were submitted to a commercial laboratory for biocompatibility testing where they passed a series of biocompatibility tests and were shown to perform at an equivalent level and with no significant differences in biocompatibility when compared with the same grade of implantable wrought alloy.

As more parts are being made from titanium alloys via MIM, the researchers are increasingly producing parts with excellent tensile properties. However, the presence of information containing the fatigue properties which could be useful for biomedical implant applications is scarce. Thus in a study that was intended to manufacture MIM titanium with improved fatigue properties, Ebel et al. [28] studied the mechanical, biological, and corrosion properties of specimens manufactured from Ti-6Al-4V-0.5B alloy. There were concerns that there may be unknown reactions in the body environment due to the boron content. Hence corrosion and biocompatibility tests were performed. The human osteosarcoma cell line MG-63 cells were cultured to perform adhesion, proliferation, and viability experiments and the results showed that the alloy Ti-6Al-4V-0.5B satisfies the requirements of a permanent implant material manufactured by MIM [28].

More studies on the suitability of MIM titanium for biomedical implant applications was carried out by Auzene et al. [44]. Their study presented results of the surface quality of MIM CP-Ti and MIM Ti-64 and its effect on biocompatibility. Comparisons were carried out between machined grade 4 CP-Ti, MIM grade 4 CP-Ti and MIM grade 4 CP-Ti with three different commercial medical implant coatings (BIOCOAT®, BIODIZE® and BIOCER®: Steiger Galvanotechnique SA., Chatel-Saint-Denis, Switzerland). Thermanox® was used a control sample. The results revealed that in comparison to machined CP-Ti, MIM CP-Ti had a specific surface roughness which exhibited an excellent biological response. Cell adhesion of the cultured bone explants was poor on the Thermanox® control, and much improved on the MIM-Ti, BIODIZE® and BIOCOAT® but did dramatically increase on BIOCER®. The chemical and mechanical properties conformed to ASTM F67 standards for MIM CP-Ti and to ASTM standards F136 and F2885 for MIM Ti-64 [44].

More results of the study mentioned above were then published by Demangel *et al.* [45]. Their report published results of the biocompatibility of MIM CP-Ti with various anodic oxidation post-treatments as reported by Auzene *et al.* [44]. It was shown that MIM-Ti compared to machined CP-Ti demonstrated a specific surface topography with a higher roughness. MIM-Ti and BIOCER® samples significantly enhanced cell proliferation, cell adhesion and cell differentiation of bone explants compared to CP-Ti. In addition the authors performed some anodisation post-treatment in this study which demonstrated the ability to improve osseointegration through anionic modification treatment. Figure 6 shows the topographic 3D view of machined CP-Ti (a) and MIM CP-Ti (b) surfaces, 1 × 1 mm studied by Demangel *et al.* [45].

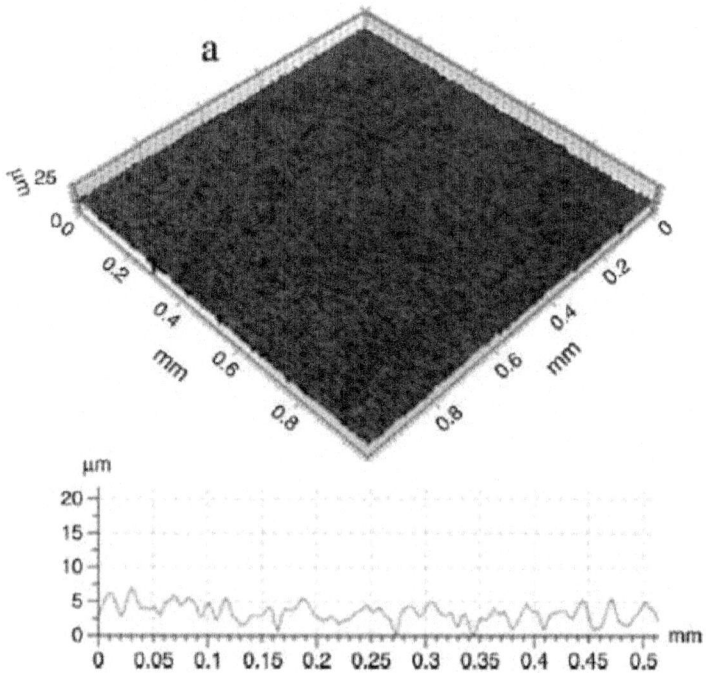

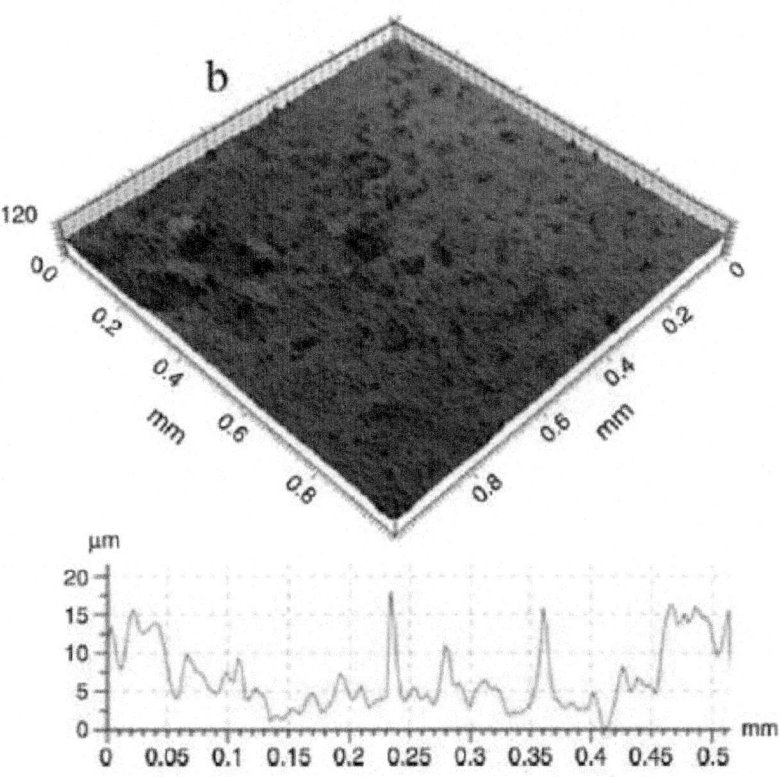

Figure 6: Topographic 3D view of machined CP-Ti (**a**) and MIM CP-Ti (**b**) surfaces, 1 × 1 mm [45].

Yet more studies relating to the study of the biocompatibility of MIM Ti-64 were carried out by Ibrahim *et al.* [46]. Firstly, the *in vitro* cytotoxicity of MIM Ti-64 was carried out using the mouse fibroblast L929 cell lines. The MIM Ti-64 part was found to be non-toxic to mouse fibroblast cell lines L929. Most of the cells proliferated with numerous filopodia and were attached to the MIM Ti-64 part [46]. Secondly, Ibrahim *et al.* [46] carried out an *in vivo* test of the MIM Ti-64 by placing the implant into the mandible of Macaca fascicularis. Results showed continuous contact between surrounding tissue and the Ti-64 implant as shown in Figure 7 [46].

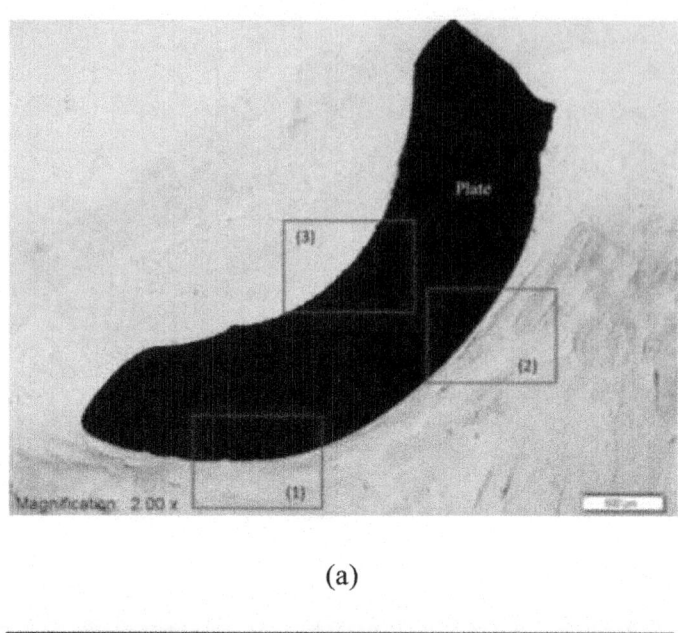

(a)

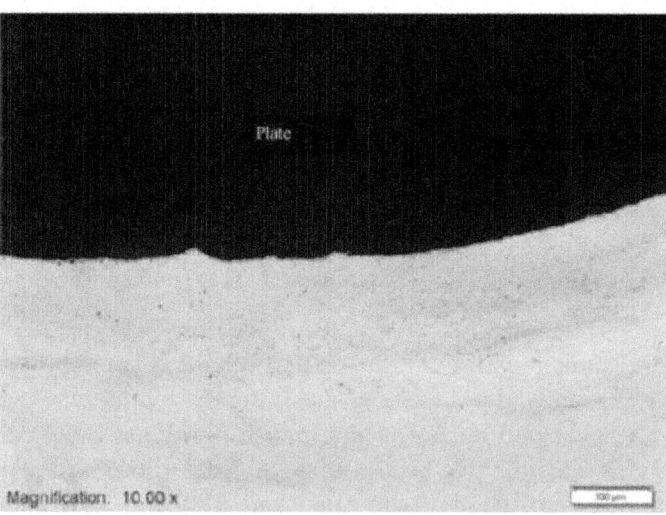

(b)

Figure 7: Micrograph showing surrounding tissue on the MIM Ti-64 implant showing continuous contact [46].

Other Advanced Manufacturing Techniques

There are a number of variations of additive manufacturing that have been employed by researchers to make implants. One such technique is Laser engineered net shaping or LENS™ which is a technology developed by Sandia National Laboratories. Similarly to EBM and DSLM, LENS™ is used for fabricating metal parts directly from a computer-aided design (CAD) solid model. The difference is that in LENS™ the metal powder is injected into a molten pool which is created by a focused, high-powered laser beam. Figure 8 is a schematic representation of LENS™ process.

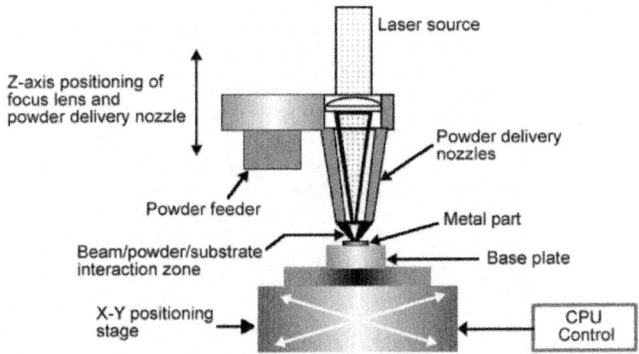

Figure 8: Schematic representation of LENS process.

LENS™ produced Ti parts have a rough surface with a macroporous structure and it is this property which allowed for Xue *et al.* [47] to use this method to fabricate porous CP-Ti implants. The effects of the porous structure on bone cell responses were evaluated *in vitro* using human osteoblast cells (OPC1). In the results the cells were found to be well spread on the surface of porous Ti and they formed strong local adhesion with assays showing that the porous CP-Ti surface favours bone cell proliferation. The porous CP-Ti LENS™ implants were also shown to stimulate faster OPC1 cell differentiation compared to a polished CP-Ti sheet. It was stated that this is due to the change in cell morphology within the pores of CP-Ti implant samples. The results also showed that a critical pore size of >200 μm is crucial for cell ingrowth into the pores. When the pore size is <150 μm, cells were found to span directly across the pores [47].

This influence of porosity on LENS™ manufactured titanium was also studied by Bandyopadhyay *et al.* [48]. After fabricating porous Ti-64 structures, using the LENS™ technology, the authors sought to demonstrate that advanced manufacturing techniques such as LENS™ can be used to fabricate low-modulus implants, with tailored porosity and which are capable of achieving long-term *in vivo* stability. At a porosity of 23–32 vol %, the corresponding effective elastic modulus was tailored between 7 and 60 GPa, equivalent to human cortical bone. The *in vivo* tests carried out on male Sprague–Dawley rats for 16 weeks revealed that there was significant increase in calcium within the implants, which is an indicator of excellent biological tissue ingrowth through interconnected porosity. Results also showed that the total amount of porosity influences the role in tissue ingrowth [47]. Figure 9 shows SEM micrographs of OPC1 cells and Ti after 3 days of culture on: (a,b) porous Ti (27% porosity), showing flattened and well-spread morphology; (c) Ti plate, showing more rounded shape.

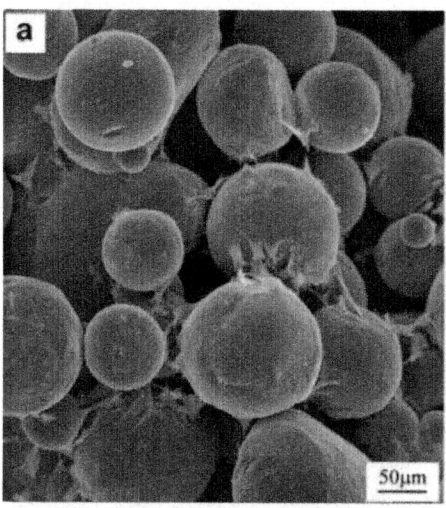

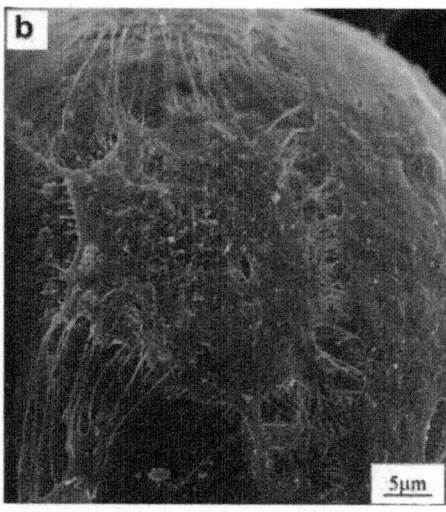

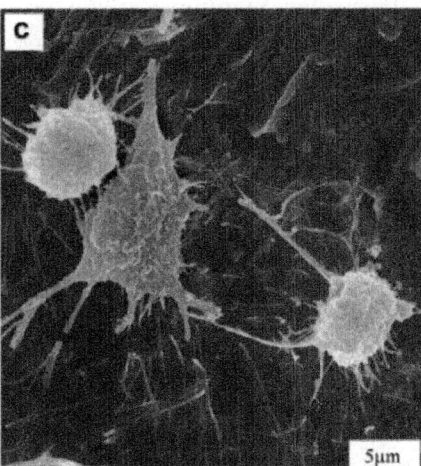

Figure 9: SEM micrographs of OPC1 cells after 3 days of culture on: (**a,b**) porous Ti (27% porosity), showing flattened and well-spread morphology; (**c**) Ti plate, showing more rounded shape [47].

In additive manufacturing, there is another technique which is widely used with polymeric materials. This method is known as fused deposition modelling (FDM). In FDM, the material is extruded in building a part layer by layer from a CAD file. The FDM schematic is shown in Figure 10. This technique also gives room for flexibility in design and this is undoubtedly beneficial for implant fabrication because implant size and shape can be tailored leading to the ability to produce patient specific implants.

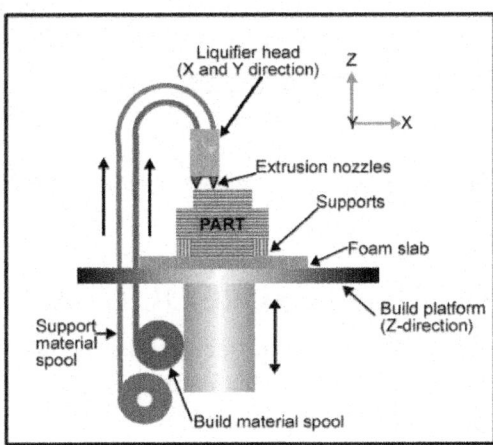

Figure 10: Fused deposition modelling schematic.

Wiria *et al.* [49] fabricated titanium implant prototypes using this 3-D printing technique in combination with the debinding and consolidation (sintering) processing that is carried out in MIM. In their study, a 3D Printer normally used for plaster material was used to fabricate green parts of the titanium implant prototype. The CP-Ti powder (325 mesh) was mixed with Poly(vinyl alcohol) (PVA) binder material to make the feedstock. After 3-D printing, the green parts were firstly debound from the PVA binder and then sintered. The subsequent cell culture study which was carried out using the L929 fibroblast cells showed that the porous CP-Ti implant had excellent biocompatibility because of the resulting bone cell attachment and proliferation. The *in vitro* investigation also revealed increased osteogenic differentiation, with little cytotoxicity. Figure 11 shows comparison of cytotoxicity between the CP-Ti scaffolds and control cylinder. The negative control samples used for cytotoxicity tests were in the form of agarose gel cylinders with the same dimension as the titanium implants, whereas phenol samples were used as positive control.

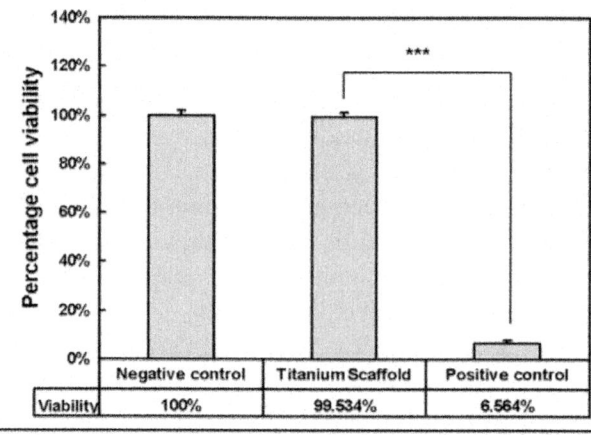

*** $p < 0.001$

Figure 11: Comparison of cytotoxicity between titanium scaffolds and control cylinder [49].

CONCLUSIONS AND FUTURE OUTLOOK

Titanium and its alloys have a proven track record as biomedical implants due to their excellent biocompatibility. This is related to the presence of an oxide layer that is typically present on the surface of the material in an oxidising environment. The processing of titanium via the advanced manufacturing technologies of ALM and MIM was until recently inhibited by relatively high costs and the need to demonstrate that the resulting surfaces have the equivalent biocompatibility of implants manufactured using traditional methods. This review has demonstrated that advanced and additive manufacturing can be used successfully to manufacture safe, biocompatible titanium alloy structures for use as medical devices in some applications. This conclusion is supported by a number of *in vitro* and *in vivo* studies.

The studies used cultured fibroblasts and osteoblasts in the observation of cell responses to surfaces and also human and animal subjects. Several advanced manufacturing routes have also been shown here to provide an enabling environment for the manufacture of implants with tailored surfaces, porosity and scaffolds. For example, this is a feature that has been used in the fabrication of titanium implants where the elastic modulus matches that of bone tissue. Table 4 and Table 5 summarise the ALM and MIM processing methods and findings.

The processing of titanium alloys manufactured via advanced powder manufacturing routes such as additive layer manufacturing (or 3D printing) and metal injection moulding is clearly receiving increased attention and being adopted as alternative to machining and casting. This is as a result of the associated key advantages which include the design flexibility, reduced processing costs, reduced waste, energy efficiency and improved functionality as has been demonstrated in several of the above studies. Therefore, the advanced manufacturing routes collectively represent a significant opportunity to transform the medical device industry. On the other hand, whilst ALM offers design flexibility, the full scale adoption in implant manufacturing may be hindered by the current high cost of machines with metal 3-D printing capabilities and by costs associated with skills training. The field of advanced manufacturing is evolving rapidly, with new technologies and discoveries appearing almost continuously and contributing to a very dynamic field. Furthermore, it has been shown that ALM can produce implants with customised rough surfaces, but in some implant cases such as in joints, a smooth surface is required and ALM and MIM are currently not capable of producing parts with very smooth surface finishes.

This means that in some applications, ALM and MIM Ti implants may have to be post-processed or coated to achieve the adequately smooth surfaces. There are also further complexities in that the cell attachment or bone ingrowth, for example, may not be uniform throughout a single Ti implant part.

Table 4: Summary of ALM processing methods and findings

Processing	Alloy	Biocompatibility test	Cell line/implantation	Other comments and references
EBM	Ti6Al4V	*In vitro*	human fetal osteoblasts (hFOB 1.19)	Reduced cell proliferation in highly rough surfaces [33]
EBM	Ti6Al4V	*In vivo*	Frontal skull of domestic pig	More bone contact in more porous samples [17]
EBM	Ti6Al4V	*In vivo*	Rabbit femur and tibia	As-EBM implant response comparable to machined [34]
EBM	Ti6Al4V	*In vitro*	Human adipose-derived adult stem cells (hASC)	Increased proliferation on porous compared to polished and unpolished EBM discs [29]
EBM	Ti6Al4V	*In vitro* and *in vivo*	Osteoblasts extracted from Calvaria of rabbits, Calvaria of rabbits	Proliferation in porous EBM Ti-64 implants matched coated implants [35]
EBM	Ti6Al4V	*In vivo*	Sheep	High bone–implant contact in porous implant [36]
DMLS	Ti6Al4V	*In vitro*	human osteoblasts cells (HOB)	Cultured cells attached and proliferated on SLM substrates [37]
DMLS	Unspecified Ti	*In vitro* and *in vivo*	BMP-7 transduced human gingival Fibroblasts	*In vitro* and *in vivo* test data showing substantial bone ingrowth [38]
DMLS	Ti6Al4V	*In vivo*	Human anterior mandible, minipig mandibular	Peri-implant bone in close contact with the surface of the implant [39]
DMLS	Ti6Al4V	*In vitro*	Human osteoblasts	Osteoblasts well-spread and with multiple contact points [40]
LENS	CP-Ti	*In vitro*	human osteoblast cells (OPC1)	Cells well spread on porous Ti [47]
LENS	Ti6Al4V	*In vivo*	Male Sprague–Dawley rats	Increase in calcium (bone) within implant pores [48]
Modified FDM	CP-Ti	*In vitro*	L929 mouse fibroblast	Excellent bone cell attachment and proliferation [49]

Table 5: Summary of MIM processing and findings

Processing	Alloy	Biocompatibility test	Cell line/implantation	Other comments and references
MIM	Ti6Al4V	*In vitro*	L929 (ISO 10993)	Passing results for ISO10993 tests [42]
MIM	Ti6Al4V0.5B	*In vitro*	MG63 cell	Alloy satisfied requirements of a MIM implant [28]
MIM	CP-Ti and Ti6Al4V	*In vitro*	MC-3T3-E1 pre-osteoblasts	Cell adhesion much improved on the MIM-Ti, BIODIZE® and BIOCOAT® [44]
MIM	CP-Ti	*In vitro*	MC-3T3-E1 pre-osteoblasts	MIM-Ti and BIOCER® had enhanced cell proliferation, adhesion and differentiation [45]
MIM	Ti6Al4V	*In vitro* and *in vivo*	L929 fibroblast and mandible of Macaca fascicularis	Cells proliferated with filopodia and attached to MIM Ti-64 [46]

Published literature clearly demonstrates that advanced powder manufacturing routes produce implants that are suitable for biomedical applications and the field of investigation of the biocompatibility and clinical performance of ALM and MIM metallic implants has been shown in this review to be expanding. However, relative to the biomedical industry and studies, the studies of *in vitro* biocompatibility, *in vivo* tissue response, and clinical performance of advanced manufactured surfaces can be considered to be limited in number. There is also room to improve the scope of these studies because of the continuing innovation in advanced manufacturing technologies. To summarise, the available data is encouraging, even where it is not always based on systematic laboratory or detailed clinical studies. There is little doubt that more research is needed into the biocompatibility and functionality of medical devices manufactured using the full range of additive and advanced manufacturing technologies available today.

ACKNOWLEDGMENTS

The author would like to thank the Wellcome Trust for providing financial support through the Institutional Strategic Support Fund (ISSF) to the University of Sheffield to conduct the present research.

REFERENCES

1. Elias, C.N.; Lima, J.H.C.; Valiev, R.; Meyers, M.A. Biomedical applications of titanium and its alloys. *JOM* 2008, *60*, 46–49.

2. BomBač, D.; Brojan, M.; Fajfar, P.; Kosel, F.; Turk, R. Review of materials in medical applications. *RMZ Mater. Geoenviron.* 2007, *54*, 471–499.

3. Mueller, E.; Kammula, R.; Marlowe, D. Regulation of biomaterials and medical devices. *MRS Bull.* 1991, *16*, 39–41.

4. Froes, F.H. Titanium powder metallurgy: A review—Part 1. *Adv. Mater. Process.* 2012, *170*, 16–22.

5. Guo, S.; Qu, X.; He, X.; Zhou, T.; Duan, B. Powder injection molding of Ti-6Al-4V alloy. *J. Mater. Process. Technol.* 2006, *173*, 310–314.

6. Sidambe, A.T.; Figueroa, I.A.; Hamilton, H.G.C.; Todd, I. Metal injection moulding of CP-Ti components for biomedical applications. *J. Mater. Process. Technol.* 2012, *212*, 1591–1597.

7. Quinn, R.K.; Armstrong, N.R. Electrochemical and surface analytical characterization of titanium and titanium hydride thin-film electrode oxidation. *J. Electrochem. Soc.* 1978, *125*, 1790–1796.

8. Schiff, N.; Grosgogeat, B.; Lissac, M.; Dalard, F. Influence of fluoride content and pH on the corrosion resistance of titanium and its alloys. *Biomaterials* 2002, *23*, 1995–2002.

9. Tengvall, P.; Lundström, I. Physico-chemical considerations of titanium as a biomaterial. *Clin. Mater.* 1992, *9*, 115–134.

10. Paschalis, E.I.; Chodosh, J.; Spurr-Michaud, S.; Cruzat, A.; Tauber, A.; Behlau, I.; Gipson, I.; Dohlman, C.H. *In vitro* and *in vivo* assessment of titanium surface modification for coloring the backplate of the boston keratoprosthesis. *Investig. Ophthalmol. Vis. Sci.* 2013, *54*, 3863–3873.

11. Adell, R.; Lekholm, U.; Rockler, B.; Brånemark, P.I. A 15-year study of osseointegrated implants in the treatment of the edentulous jaw. *Int. J. Oral Maxillof.* 1981, *10*, 387–416.

12. *Standard Specification for Unalloyed Titanium for Surgical Implant Applications (UNS R50250, UNS R50400, UNS R50550, UNS R50700)*; ASTM F67–13; American Society for Testing Materials: West Conshohocken, PA, USA, 2013.

13. *Standard Specification for Wrought Titanium-6 Aluminum-4 Vanadium ELI (Extra Low Interstitial) Alloy for Surgical Implant Applications (UNS R56401)*; ASTM F136–13; American Society for Testing Materials: West Conshohocken, PA, USA, 2013.

14. Sidambe, A.T.; Choong, W.L.; Hamilton, H.G.C.; Todd, I. Correlation of metal injection moulded Ti6Al4V yield strength with resonance frequency (PCRT) measurements. *Mater. Sci. Eng. A Struct.* 2013, *568*, 220–227.

15. Zhao, X.; Chen, L.; Xin, L.; Huang, W. Study on microstructure and mechanical properties of laser rapid forming Inconel 718. *Mater. Sci. Eng. A Struct.* 2008, *478*, 119–124.

16. Bidaux, J.E.; Closuit, C.; Rodriguez-Arbaizar, M.; Zufferey, D.; Carreno-Morelli, E. Metal injection moulding of low modulus Ti-Nb alloys for biomedical applications. *Powder Metall.* 2013, *56*, 263–266.

17. Ponader, S.; von Wilmowsky, C.; Widenmayer, M.; Lutz, R.; Heinl, P.; Körner, C.; Singer, R.F.; Nkenke, E.; Neukam, F.W.; Schlegel, K.A. *In vivo* performance of selective electron beam-melted Ti-6Al-4V structures. *J. Bio. Mater. Res. A* 2009, *92*, 56–62.

18. Parthasarathy, J.; Starly, B.; Raman, S.; Christensen, A. Mechanical evaluation of porous titanium (Ti6Al4V) structures with electron beam melting (EBM). *J. Mech. Behav. Biomed. Mater.* 2010, *3*, 249–259.

19. Murr, L.E.; Amato, K.N.; Li, S.J.; Tian, Y.X.; Cheng, X.Y.; Gaytan, S.M.; Martinez, E.; Shindo, P.W.; Medina, F.; Wicker, R.B. Microstructure and mechanical properties of open-cellular biomaterials prototypes for total knee replacement implants fabricated by electron beam melting. *J. Mech. Behav. Biomed. Mater.* 2011, *4*, 1396–1411.

20. *Standard Specification for Metal Injection Molded Titanium-6Aluminum-4Vanadium Components for Surgical Implant Applications*; ASTM F2885–11; American Society for Testing Materials: West Conshohocken, PA, USA, 2011.

21. *Standard Specification for Additive Manufacturing Titanium-6 Aluminum-4 Vanadium ELI with Powder Bed Fusion*; ASTM F2924–14; American Society for Testing Materials: West Conshohocken, PA, USA, 2014.

22. *Standard Specification for Additive Manufacturing Titanium-6 Aluminum-4 Vanadium ELI (Extra Low Interstitial) with Powder Bed Fusion*; ASTM F3001–14; American Society for Testing Materials: West Conshohocken, PA, USA, 2014.

23. *Sintered-Metal Injection-Moulded Materials—Specifications*; ISO 22068-14; International Organisation for Standardisation ISO: Geneva, Switzerland, 2014.

24. Sidambe, A.T.; Figueroa, I.A.; Hamilton, H.G.C.; Todd, I. Taguchi optimization of MIM titanium sintering. *Int. J. Powder Metall.* 2011, *47*, 21–28.

25. Froes, F.H. Titanium powder metallurgy: A review—Part 2. *Adv. Mater. Process.* 2012, *170*, 26–29.

26. Van Noort, R. The future of dental devices is digital. *Dent. Mater.* 2012, *28*, 3–12.

27. Ferri, O.M.; Ebel, T.; Bormann, R. High cycle fatigue behaviour of Ti-6Al-4V fabricated by metal injection moulding technology. *Mat. Sci. Eng. A Struct.* 2009, *504*, 107–113.

28. Ebel, T.; Blawert, C.; Willumeit, R.; Luthringer, B.J.C.; Ferri, O.M.; Feyerabend, F. Ti-6Al-4V-0.5B: A modified alloy for implants produced by injection molding. *Adv. Eng. Mater.* 2011, *13*, B440–B453.

29. Haslauer, C.M.; Springer, J.C.; Harrysson, O.L.A.; Loboa, E.G.; Monteiro-Riviere, N.A.; Marcellin-Little, D.J. *In vitro* biocompatibility of titanium alloy discs made using direct metal fabrication. *Med. Eng. Phys.* 2010, *32*, 645–652.

30. Chen, H.; Sago, A.; West, S.; Farina, J.; Eckert, J.; Broadley, M. Biocompatibility of metal injection molded *versus* wrought ASTM F562 (MP35N) and ASTM F1537 (CCM) cobalt alloys. *BioMed. Mater. Eng.* 2011, *21*, 1–7.

31. Ivanova, O.; Williams, C.; Campbell, T. Additive manufacturing (AM) and nanotechnology: Promises and challenges. *Rapid Prototyp. J.* 2013, *19*, 353–364.

32. Al-Bermani, S.S.; Blackmore, M.L.; Zhang, W.; Todd, I. The origin of microstructural diversity, texture, and mechanical properties in electron beam melted Ti-6Al-4V. *Metall. Mater. Trans. A* 2010, *41A*, 3422–3434.

33. Ponader, S.; Vairaktaris, E.; Heinl, P.; Wilmowsky, C.V.; Rottmair, A.; Körner, C.; Singer, R.F.; Holst, S.; Schlegel, K.A.; Neukam, F.W.; *et al.* Effects of topographical surface modifications of electron beam melted Ti-6Al-4V titanium on human fetal osteoblasts. *J. Biomed. Mater. Res. A* 2008, *84A*, 1111–1119.

34. Thomsen, P.; Malmström, J.; Emanuelsson, L.; René, M.; Snis, A. Electron beam-melted, free-form-fabricated titanium alloy implants: Material surface characterization and early bone response in rabbits. *J. Biomed. Mater. Res. B* 2009, *90B*, 35–44.

35. Li, X.; Feng, Y.F.; Wang, C.T.; Li, G.C.; Lei, W.; Zhang, Z.Y.; Wang, L. Evaluation of biological properties of electron beam melted Ti6Al4V implant with biomimetic coating *in vitro* and *in vivo*. *PLos One* 2012, *7*, e52049.

36. Palmquist, A.; Snis, A.; Emanuelsson, L.; Browne, M.; Thomsen, P. Long-term biocompatibility and osseointegration of electron beam melted, free-form-fabricated solid and porous titanium alloy: Experimental studies in sheep. *J. BioMater. Appl.* 2013, *27*, 1003–1016.

37. Hollander, D.A.; Wirtz, T.; Walter, M.V.; Linker, R.; Schultheis, A.; Paar, O. Development of individual three-dimensional bone substitutes using "selective laser melting". *Eur. J. Trauma* 2003, *4*, 228–234.

38. Hollister, S.J.; Lin, C.Y.; Saito, E.; Lin, C.Y.; Schek, R.D.; Taboas, J.M.; Williams, J.M.; Partee, B.; Flanagan, C.L.; Diggs, A.; *et al.* Engineering craniofacial scaffolds. *Orthod. Craniofac. Res.* 2005, *8*, 162–173.

39. Mangano, C.; Piattelli, A.; d'Avila, S.; Iezzi, G.; Mangano, F.; Onuma, T.; Shibli, J.A. Early human bone response to laser metal sintering surface topography: A histologic report. *J. Oral Implantol.* 2010, *36*, 91–96.

40. Warnke, P.H.; Douglas, T.; Wollny, P.; Sherry, E.; Steiner, M.; Galonska, S.; Becker, S.T.; Springer, I.N.; Wiltfang, J.; Sivananthan, S. Rapid prototyping: Porous titanium alloy scaffolds produced by selective laser melting for bone tissue engineering. *Tissue Eng. Part C Methods* 2009, *15*, 115–124.

41. Todd, I.; Sidambe, A.T. Developments in metal injection moulding (MIM). In *Advances in Powder Metallurgy: Properties, Processing and Applications*; Isaac Chang, Y.Z., Ed.; Woodhead Publishing Limited: Sawston, UK, 2013; pp. 109–146.

42. Sago, J.A.; Broadley, M.W.; Eckert, J.K.; Chen, H. Manufacturing of implantable biomedical devices by metal injection moulding. *Adv. Powder Metall. Part Mater.* 2010, *4*, 89–99.

43. Sago, J.A.; Broadley, M.W.; Eckert, J.K. Metal injection molding of alloys for implantable medical devices. 2012,*48*, 41–49.

44. Auzene, D.; Mallejac, C.; Demangel, C.; Lebel, F.; Duval, J.L.; Vigneron, P.; Puippe, J.C. Influence of surface aspects and properties of MIM titanium alloys for medical applications. *PIM Int.* 2012, *6*, 57–61.

45. Demangel, C.; Auzène, D.; Vayssade, M.; Duval, J.-L.; Vigneron, P.; Nagel, M.-D.; Puippe, J.-C. Cytocompatibility of titanium metal injection molding with various anodic oxidation post-treatments. *Mater Sci. Eng. C* 2012, *32*, 1919–1925.

46. Ibrahim, R.; Azmirruddin, M.; Jabir, M.; Muhamad, N.; Rafiq, M.; Hayaty, N.; Kasim, A.; Muhamad, S.; Hanada, K.; Shimizu, T.; *et al.* *Pre-Clinical Study on the Oral Maxillofacial (OMF) Titanium Alloy Implants Produced By Metal Injection Molding (MIM) Using Palm Oil Based Binder System*; Euro PM Congress and Exhibition and EPMA: Gothenburg, Sweden, 2013.

47. Xue, W.; Krishna, B.V.; Bandyopadhyay, A.; Bose, S. Processing and biocompatibility evaluation of laser processed porous titanium. *Acta Biomater.* 2007, *3*, 1007–1018.

48. Bandyopadhyay, A.; Espana, F.; Balla, V.K.; Bose, S.; Ohgami, Y.; Davies, N.M. Influence of porosity on mechanical properties and *in vivo* response of Ti6Al4V implants. *Acta Biomater.* 2010, *6*, 1640–1648.

49. Wiria, F.E.; Shyan, J.Y.M.; Lim, P.N.; Wen, F.G.C.; Yeo, J.F.; Cao, T. Printing of titanium implant prototype.*Mater. Des.* 2010, *31*, S101–S105.

Chapter 10

LINKING PROCESS TECHNOLOGY AND MANUFACTURING PERFORMANCE UNDER THE FRAMEWORK OF MANUFACTURING STRATEGY

Hongyi Sun

Department of Systems Engineering and Engineering Management City University of Hong Kong China

INTRODUCTION

Performance improvement is the goal of any manufacturing firms. A bunch of manufacturing practices are involved as suggested in the manufacturing strategy model. These include technologies, human resources and comprehensive programmes such as total quality management (TQM) and pull production. As a result, the linkage among various practices and performance are very complicated. Previous research in this field may have some limitations. The following part will review these limitations using TQM and AMT (Advanced Manufacturing Technology) as an example and argue the necessity of using structural equation modeling to deal with multiple variables.

First, most previous research on practice-performance linkage assumes that all practices directly contribute to the performance. Therefore, the conceptual models are mostly a one-layer model. The data analysis methods are mostly simple correlation or multiple correlation. The methodology is basically exploratory. The assumption of this research argues that practices may not all be directly correlated with performance. There may be several layers from practices to performance. Therefore, a comprehensive model based on path analysis or structural equation modeling is needed to investigate the practice-performance relationship. To specify the path-analysis model, a conceptual model is needed. In this research the conceptual framework from manufacturing strategy will be used.

Second, in previous research, the measures of practices vary from one single question to a set of questions which are grouped into a construct. It is not so common to develop constructs in AMT-performance research yet. The definition and classification of AMT are not consistent. Beaumont et al (2002) measure AMT in terms of direct (fabrication and assembly), indirect (engineering and design) and administrative (information management). Dasa and Narasimhan (2001) divided AMT into manufacturing technologies and design technologies. However, the classification of AMT is not consistent with technical definition (Groover, 1987; Goetsch, 1990; Singh, 1996; Kotha and Swamidass, 1998). In this research, AMT will be classified according to technical definition of computer integrated manufacturing (CIM).

In summary, practice-performance linkage has been mostly studied by simple or multiple correlation analysis in single areas such as technology or quality. In modern manufacturing companies, both practices as input and manufacturing performance as output are getting more and more complicated. Therefore, the relationship must be a complex one. This paper reports the research which aims to investigate this complex practice-performance linkage in a path-analysis model. The research is based on the manufacturing strategy framework. The idea is consistent with complex performance. Complex performance is described by Lewis and Roehrich (2009) in terms of the interaction between infrastructural complexity (e.g. buildings, enabling facilities, hardware) and transactional complexity (e.g. performance involving high degrees of embedded knowledge).

The paper is structure in five sections. In section two, literature on all types of practice and performance will be reviewed under the framework of manufacturing strategy and a set of hypotheses will be formulated. In section three, methodological issues such as data collection, operationalisation, validity and reliability tests and data analysis method will be described. In section four, the results will be presented. In section five, the results will be discussed and implications for practice and future research will be explored. In the final section, the research will be concluded; limitation and future research will be discussed.

A CONCEPTUAL MODEL AND HYPOTHESES FORMULATION

The Conceptual Framework under Manufacturing Strategy

Manufacturing strategy is regarded as the manner in which the business unit deploys its manufacturing resources (Hayes and Wheelwright, 1984) and effectively uses its manufacturing strengths (Swamidass and Newell, 1987; Riis, 1992) to complement the business strategy. One of the themes in manufacturing strategy deals with various linkages or alignment among business objectives, manufacturing missions, manufacturing practices and performance. This paper aims to explore the relationship between manufacturing practices and performance. The key variables are practices and performance. The related variables include performance, structural decisions, infrastructural decisions, technology, and organization. The contents and possible relationships among the variables are illustrated in figure 1 and will be elaborated below.

Manufacturing Performance

Under manufacturing strategy theory, manufacturing practices may not directly contribute to business performance such as market share and profitability. Their immediate contribution should be those at manufacturing levels such as cost reduction, quality improvement and shortening throughput time. Therefore, in manufacturing strategy research, business performance and manufacturing performance are distinguished (Tunalv 1991, McDermott and Stock, 1999, Sun and Cui 2002, Beaumont et al, 2002). These manufacturing performance dimensions, if being well aligned with business competitive objectives, will contribute to the achievement of business performance (Dasa and Narasimhan, 2001, Sun and Cui 2002). Therefore, there should be a corresponding relationship between manufacturing performance, manufacturing missions and business objectives. So in this research on practice-performance linkage, the performances refer to manufacturing performance. In manufacturing strategy research, manufacturing performance should be corresponding to manufacturing missions/tasks which cover cost, quality, delivery, flexibility and service (Skinner 1969, Wheelwright 1984, Kim and Arnold 1996 etc.). The service often refers to customer satisfaction. Based on the alignment and corresponding theory, manufacturing performance can also be divided on into these five categories.

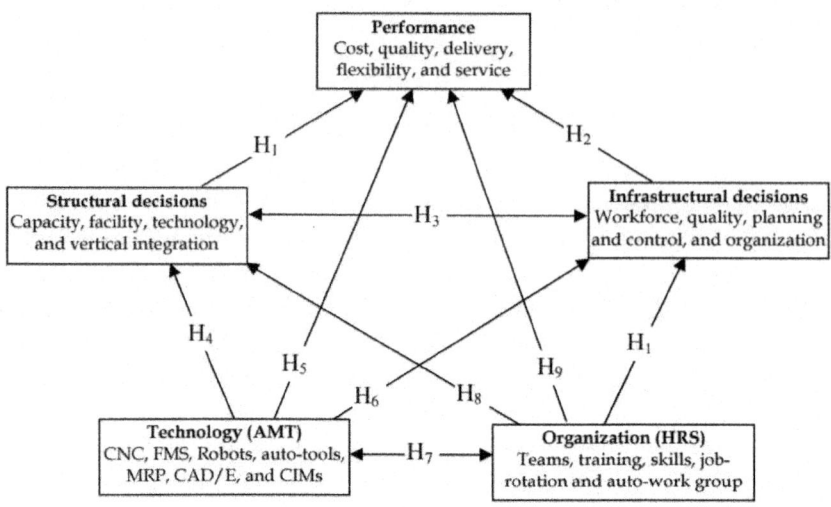

Figure 1: A conceptual model for studying practice-performance linkage

Action Programmes Based On Structural and Infrastructural Decisions

Manufacturing action programs are often regarded as sets of decisions, that derive from the experience of a number of leading companies and that have proved to be successful (Schonberger, 1982; 1986; Hanson and Voss, 1993, Hanson et al., 1994). They are the resources or functions that must be performed by manufacturing (Schroeder et al., 1986). Because of the diversity of manufacturing decisions that must be made over time, Hayes and Wheelwright (1984) developed an organizing framework that groups them into two major categories, structural and infrastructural decisions. There is an essential agreement on this structure-infrastructure dichotomy in the literature (e.g., Leong et al., 1990; Hill 1995, Tseng et al, 1999, Ng and Hung 2001). Structural decision category addresses the "bricks and mortar" decisions of capital spending. Examples of structural decisions include decisions on capacity, facility, the investment in technology, and vertical integration (Hayes and Wheelwright, 1984). Infrastructural decision category addresses more "tactical" issues, which affect the people and systems that make manufacturing work (Leong et al., 1990). The infrastructural decisions may include decisions on workforce, quality, production planning and organization Corresponding to the above two decision areas, there are two types of action programmes. Those programmes supporting structural decisions such as increasing equipment and capacity are named as structural programmes. The programmes to support infrastructural decisions and choices are named as infrastructural programmes. Regarding contribution to performance, Hayes *et al.* (1988) suggested that infrastructure decisions were equally important as structure decisions. Performance improvement has been found positively correlated with infrastructural programs such as quality management programs, pull production systems, total productive maintenance (Cua, McKone and Schroeder, 2001), and supply chain management. Structural and infrastructural decisions are the two sides of the same manufacturing process. So they must be related to each other. Hayes *et al.* (1988) suggested that the distinction between structure and infrastructure was analogous to the distinction between computer hardware and software. The fixed, long-term and often unrecoverable investments of the firm in durable or facilities are analogous to computer hardware, while those that are more controllable by management are analogous to software. Based the contents and analysis of the two types of programmes, the following hypotheses are formulated.

H_1: Structural programs directly contribute to Performance.

H_2: Infrastructural programs directly contribute to Performance.

H_3: Infrastructural programs are positively related to Structural programs.

Technology

Under manufacturing strategy framework, technology is part of structural decisions. However, since technology has changed dramatically in the past decades years and it has very different features compared with other items such as capacity and facility etc, technology is treated separately and refers to Advanced Manufacturing Technologies (AMT).

AMT refers to those computer-aided technologies in information management, design, engineering and fabrication processes such as Computer Aided Manufacturing (CAM), Computer Aided Design (CAD) and Computer Aided Process Planning (CAPP). AMTs are the main technical components of Computer Integrated Manufacturing (CIM) systems. It is more than a group of advanced and automated technologies (Haywood, 1990). The main feature of CIM is the total integration of all manufacturing functions, including design, engineering, planning, control, fabrication, and assembly etc. through the use of computers. According to the CIM wheel model of the Society of Manufacturing Engineer (SME), there are one business and four technical components of a CIM system (Goetsch 1990). The four technical components are *planning and controlling, information resources management, product and process definition,* and *factory automation.* The four components and relevant AMTs involved have been described in details in literature (Groover, 1987;

Goetsch, 1990; Singh, 1996; Kotha and Swamidass, 1998). The contents of the four components as well as their relationship with other variables will be analyzed below.

The factory automation component contains will directly influence the structural decision on the manufacturing process, especially the level of automation, new equipment implementation, capacity incensement and facility investment (Goetsch 1990, Bessant and Haywood 1988). In fact, the structural decision is called process choices in Hill's model (Hill 1995). Regarding the relationship between processes and AMT selection, there have been many similar models reported (Fix-Sterz et al 1987, p.11, Greenwood 1988, Lindberg 1990 p.12, Noori 1990, Ayres 1991, Parthasarthy & Sethi 1992). In general, for small batch and large variety job shop processes, standalone NC and MC will be suggested. For medium batch and variety, FMS is recommended. For large volume and few varieties, dedicated and automated lines are suggested. All these suggest that different processes may use different type of AMTs. In either case, the changes in process will require the changes in the technological dimensions. In other words, AMT is needed to support the implementation of structural programs for the purpose of updating manufacturing processes. The above reference leads to the fourth hypothesis.

H_4: The implementation of structural programs will be positively correlated with the utilization of manufacturing technologies.

The planning and controlling component includes such elements as planning/scheduling and controlling of facilities, materials, tools and shop floor activities. Hardware and software are available to automate each of the elements. Material Requirement Planning (MRP), as well as Manufacturing Resources Planning (MRP II), is an important concept with a direct relationship to CIM. Information resources management is the nucleus of CIM. Information, updated continually and shared instantaneously, is what CIM is all about. One of the major goals of this nucleus is to overcome the barriers that prevent the complete sharing of information among all other CIM components. The AMTs used for this purpose include Shared Databases (Shared DB), Wide Area Network (WAN), and Local Area Network (LAN). Planning and control is one of the key issues in infrastructural decision. However, it needs the support of technologies such as MRP and IT system. Re-engineering program is especially based on IT system implementation. The implementation of IT systems also needs the support of the relevant infrastructural changes. The above analysis leads to H_5.

H_5: The utilization of AMT is positively correlated with the implementation of infrastructural programs.

The need to achieve cost efficiency, quality, and flexibility is necessary, and has imposed a major challenge to the manufacturing industry in the nineties and beyond. AMT has been widely regarded as a new and valuable weapon to rise to the challenge proposed by the new market situation to manufacturing industries (Hunt, 1987; Noori, 1990). Therefore, AMT is widely regarded as the new weapon to improve manufacturing performance. This leads to the following hypothesis.

H_6: There is a positive relationship between technology utilization and manufacturing performance.

Organizational Dimension

Workforce and organisation are part of the infrastructural decisions. However, the issue is different to other items such as quality, planning and control. Additionally, HRS and organisational issues have been studied intensively from AMT perspective. So the organisational issue is separated in the research. Since the scope of study is in manufacturing function, the organisation refers to work organisation on the shop floor.

Plenty of previous research was found on the changes in human resources in association with single AMTs. Lee and Leonard (1990) discovered that the Automated Guided Vehicle (AGV) in a small batch-manufacturing environment altered the nature of human work. Saraph and Sebastian (1992) reviewed many previous studies and concluded that the failure of AMT is mainly due to the implicit or explicit neglect of critical human resource factors. Gerwin and Kolandy (1992, p.215) said that AMT invites a wide range of changes in human resources management and practices. They further suggested that human resources development should be integrated with the design of new technologies in the manufacturing environment. Samson, Sohal and Ramsay (1993) argue that human resources issues such as commitment, involvement, the acceptance of changes, culture, work and skills should be considered for the successful implementation of AMT. According to these previous studies, the human resources suitable for AMT are characterised by lower division of labour, frequent job rotation, stable employment, active employees' participation, loose first-line supervision, more training, team-based work organisation, group-based incentive system (Sun 2001). Based on the requirement of the development in HRS and organisational dimension for AMT implement, the following hypotheses can be formulated.

H_7: The utilization of AMT is positively correlated with the adoption of new form of work organization.

The most influential research on organizational structure and technology was made by Woodward (1965) at Imperial College in England. The very original research was conducted through a survey of 203 British manufacturing firms (p.8). Woodward's research was carried out at the level of the work organization in the production department. The samples are purely industrial companies. Woodward found that type of production, i.e., the structural decision area, was related to a specific type of organizational structure. The found that production process was the most important factor deciding the organizational structure. The number of levels in the management hierarchy, the span of control of first-line supervisors, and the ratio of managers and supervisors to other personnel were all affected by the type of the employed production technology. Besides, the success or effectiveness of the organizations was related to the "fit" between processes and organizational structure. The successful firms of each type were those that had the appropriate structured technical systems. The theory leads to hypothesis H9.

H_8: The adoption of new work organization is positively correlated to the implementation of structural action programs.

HRS and organization is part if infrastructural decision area, there it is of course related to the infrastructural decisions and relevant action programmes to support the decision. For example, teams work, employee involvement and suggestions have been proved to be a necessary part of quality management program. Employee involvement in terms of suggestions and participation are associated with quality management activities such as quality circles and communication. Research has shown that job enrichment and task characteristics such as skill variety and autonomy are directly associated with higher work quality and employee satisfaction (Kopelman, 1986). Self-managing work teams typically produce positive results in terms of quality and costs (Beekun, 1989; Sundstrom, 1990). Teams are also proved to be useful for new product development (Sobek II et al, 1998). Therefore, it is natural to formulate hypothesis H_9 and H_{10}.

H_9: The autonomous working organization is positively correlated with performance.

H_{10}: The adoption of autonomous working organization is positively correlated with the implementation of infrastructural programmers.

The relevant variables and would-be relationships are illustrated in the conceptual model as shown in figure 1. The ten hypotheses will be tested in several models.

EMPIRICAL DATA

Questionnaire and data collection

The data for this research are from the International Manufacturing Strategy Survey (IMSS). The project was initiated by London Business School and Charlmes University of Technology in 1992. IMSS is an international research network consisting of 20 countries and 600 companies around the world, including developed countries, i.e. USA, Japan, British, Germany, and developing countries, i.e. China, Argentina, Mexico. The participant companies are in the metal products, machinery and equipment industry, i.e. the international Standard Industry Classification (ISIC) 38. For details regarding IMSS project, please refer to the book by Lindberg et al., 1998.

The research reported in this paper is based on the data from the third round of IMSS survey. Data collection methods varied from country to country. In some countries, sample selection was at the coordinators' convenience, and others used random sampling. Phone contact was followed in most of the participating countries, except for the Netherlands. The questionnaires were forwarded to participating companies via mailing, fax or on-site interview. In those countries where English is not used, the questionnaire was translated into local native languages. Participating countries sent their data to the coordinator who forwarded the final database to all participants. When this research is conducted, 282 sets of data are available.

IMSS questionnaire covers four aspects of manufacturing practices and strategies. In this research those questions that are related to practice and performance are selected. In the practice part, there are three sections, namely, technology, organization and improvement programs. The section on organization contains questions on suggestions, training, skills, teams, and job rotations. The performance section contains questions related to quality, flexibility, delivery, cost and customer satisfaction. These questions are listed in the Appendix.

Method for validity and reliability tests

Validity and reliability tests cover content validity, construct validity and reliability. *Content validity* refers to whether the items in a scale represent the contents of a theoretical construct. The content validity is based on literature review, research experiences, and case studies. The contents of technology, organization, improvement programmes and performance have all been reviewed and discussed in literature review section.

Reliability refers to the internal consistent of the items within a scale that aims to measure a theoretical construct. The most commonly used test method is internal consistency (Saraph, Benson, and Schoeder, 1989; Flynn, Schroeder and Sakakibala, 1994; Nunnally, 1978). It is estimated by using Cronbach's alpha. Peterson's (1994) summary of Cronbach coefficient shows that a value above 0.7 was thought to be sufficient in most of the situations. However, in the early stage of a research where the construct had not been well tested in previous studies, Nunnally (1967) recommended a level above 0.5 be acceptable.

Construct validity refers to whether a scale is an appropriate operational definition of an abstract variable or a construct (Nunnally 1978). It is established through the use of principal factor analysis. Factor analysis (de Vaus 1993) groups variables (i.e., single questions) into factors based on their common correlation. Those variables that are correlated with each other will be grouped together. Such a group of variables is called a factor. The grouping is based on the rotated loading coefficients. The threshold of the loading coefficients is related to the size of the sample. For example, Flynn, Schroeder and Sakakibala (1994) claim that for a sample of 100, the loading of 0.19 and 0.26 indicate significance at the 0.05 and 0.01 levels, respectively. This is based on the seminal work by Cohen (1988), who suggested that in 'soft' behavioral and management research, an effect size of 0.3 is often encountered (p.95). Based on Cohen's argument, de Vaus (1993) suggested a rule of thumb as follows: if its rotated loading coefficient is more than 0.30, then a variable will be included in the corresponding factor; if the loading coefficients for all the factors are more than 0.3, then the variable will be grouped according to the largest coefficient and conceptual analysis. As the sample size of this study is 250 (180 plus 71), with a 95% confidence level and an effect size of 0.3, the statistical power of this sample is larger than 0.95 (Cohen, 1988, p.102), which is high enough to identify inherent statistical relationships.

Construct measurement

All the questions used in this research are coded and corresponding to the questionnaire in the appendix.

Manufacturing performance and the latent variable

Manufacturing performance is directly measured by asking the respondents to indicate the amount of change of the performance dimensions over the past three years, with 1=strongly deteriorated and 5=strongly improved. According to the classification of manufacturing mission and performance under manufacturing strategy, five constructs/dimensions are formulated as shown in table 1. All the constructs passed validity and reliability tests. Additionally, a second level factor analysis of the five performance dimensions produces a valid and reliable performance scale. This means that a latent variable of performance exists.

Technology constructs and the latent variables

Based on the classification in literature, AMT is divided onto four constructs, namely, fabrication (NC, MC and FMS) assembly, design (CAD/E), information technology (IT) and integrated manufacturing with automated materials transportation and inspection.

Table 1: Manufacturing performance constructs

Code	Factors and items	1	2	3	4	5	Performance
	1. Quality:						0.64
D21	Manufacturing conformance	0.74					
D22	Product quality and reliability	0.72					
	2. Flexibility:						0.90
D24	Volume flexibility		0.88				
D25	Mix flexibility		0.63				
	3. Delivery:						0.78
D28	Delivery speed			0.73			
D29	Delivery reliability			0.88			
	4. Cost:						0.55
D213	Labor productivity				0.67		
D214	Inventory turnover				0.72		
D215	Capacity utilisation				0.62		
D27	5. Service (customer satisfaction):					\	0.69
	Extraction Sums of squared Loadings						
	Total	1.73	1.64	2.67	1.88	\	1.9
	% of Variance	38.64	54.23	43.46	39.42	\	40
	(Cronbach's α)	0.70	0.77	0.77	0.63	\	0.60

dimensions produces a valid and reliable technology scale. This implies that there exist a internal validity of technology.

Confirmative factor analysis revealed that the FMS and NC, MC are separated into two factors which are named standalone automation and FMS, respectively. Other items passed the factor analysis. Finally five AMT constructs are identified. Their validity and reliability tests are list in table 2. Additionally, a second level factor analysis of the five technological

Table 2: Factor analysis of technologies by CFA

Factors and items	1	2	3	4	5
1. Integrated manufacturing:					
BT15 Robots	.712				
BT16 Automated guided vehicles (AGVs)	.602				
BT17 Automated storage-retrieval systems (AS/RS)	.721				
BT19 Computer-aided in inspecting/testing/ tracking	.666				
2. CAD/E:					
BT110 CAD; CAE		.817			
BT111 CAD-CAE-CAM-CAPP		.807			
BT112 Eng'g DB, Product Data Management systems		.654			
3. IT and MRP:					
BT23 Purchasing and supply management			.884		
BT21 Material management			.867		
BT22 Production planning and control			.786		
BT24 Sales and distribution management			.760		
BT25 Accounting and finance			.730		
BT113 LAN-WAN/ Intranet / Shared databases/Internet			.551		
4. Standalone automation:					
BT13 CNC-DNC				0.80	
BT12 Machining centers				0.77	
BT14 Automated tool change - parts loading/unloading				0.75	
BT11 Stand-alone/NC machines				0.66	
5. FMS:					0.63
Extraction Sums of Squared Loadings					
Total	1.83	1.75	3.66	2.23	/
% of Variance	45.79	58.25	59.43	55.76	/
% of Variance	45.79	58.25	59.43	55.76	/
(Cronbach's α)	0.60	0.64	0.86	0.74	/

Organsiation construct and a representative partial model

The organization part contains ten questions. Some of them were deleted since they are not relevant. Corresponding to literature review on HRS development, questions on training, skills, working in teams and job rotation are selected. Since the constructs for HRS development as discussed in the paper are not as common as AMT constructs, explorative factor analysis is used to explore all the items. It is found that the two questions related to training do not significantly related to other items. Scanning the data revealed that the data on training may have something wrong. Maybe due to different training systems, there are quite many data that are not explainable at all. For example, annual training hours are more than 10,000 hours. So questions on training are neglected. A question on labour union cooperation is also deleted since it is not a common question for all participating countries. The rest questions are analyzed and produce 3 factors which are named, working in teams, autonomous working group and suggestions, and skills and job rotation. The validity and reliability tests are shown in table 3. The construct "auto work org. & suggestions" does not pass the reliability test. Its Cronbach alphas is only 0.39, less than the minimum threshold of 0.05. The construct is not accepted. Instead, the two items "auto work org." and "suggestions" are treated as separate variables. So there are four variables in transitional dimension, namely, autonomous working organization, suggestions, working in team and skills and rotation.

Table 3: Factor analysis of human resources items by EFC

Code		F1	F2	F3
		Team	Skills & rotation	Auto work org. & suggestions
B06a	Team in fabrication	0.90	0.15	0.06
BO6b	Team in assembly	0.90	-0.01	0.12
BO9	Multiple skills	0.03	0.86	0.12
BO10	Job rotation	0.10	0.86	-0.01
BO5	Suggestions	-0.06	0.00	0.87
C512a	Auto work org.	0.30	0.13	0.67
	Rotation Sums of Squared Loadings:			
	Total	1.72	1.51	1.24
	% of Variance	28.59	25.10	20.60
	Cumulative %	28.59	53.69	74.30
	Cronbach's α	0.80	0.67	0.39<0.5

Note: ** significant at the level of $p=0.01$, * significant at the level of $p=0.05$

The second level factor analysis of the four variables does not produce a valid and reliable scale. Therefore these four factors cannot be treated as a latent variable in data analysis. Based on the correlation analysis, it is found that "autonomous working organization" is correlated with all other three variables and no other correlation relationships exist. So this variable will be used as a representative variable of organizational dimension while other three are linked to the representative one. In fact, the measure of autonomous working organization is a quite representative since it covers knowledge of employees, delegation, training, improvement and autonomous teams. Details will be shown it the specified models in figure 2, 3, and 4.

Structural and infrastructural programmes

The programmes used in this research refer to a major project aimed at producing considerable changes in the company's management practice and organization. There are fourteen improvement action programs listed in the questionnaire. These programmes cover many aspects of manufacturing improvement. However, based on manufacturing strategy framework, improvement activities can be divided into structural and infrastructural areas. Based on this concept, the programmes are divided into two groups, namely structural and infrastructural programmes as shown in table 4. These two groups of programmes both pass the validity and reliability tests as shown in table 4. This indicates that companies do no implement action programme individually, rather in a coherent and systematic way. The validity and reliability tests imply that there exist a latent variable of structural programs and a latent variable of infrastructural programs.

Table 4: Factor analysis of action programmes by CFA

Code		Component	Component
	Structural programmes:		
C53A	Process automation	.767	
C51A	Updating process equipment	.763	
C511A	Equipment productivity	.667	
C58A	Process focus	.634	
C52A	Expanding manufacturing capacity	.528	
	Infrastructural programmes:		
C59A	Pull production		.717
C513A	New product development		.713
C510A	Quality improvement		.687
C56A	Restructuring supply strategy		.623
C57A	Outsourcing		.582
C514A	Environmental compatibility		.490
	Extraction Sums of Squared Loadings:		
	Total	2.296	2.46
	% of Variance	45.928	41.06
	Cumulative %	45.928	41.06
	Cronbach's α	0.69	0.77

Structural Equation Modeling (SEM) and model fitness test

In this study, structural equation modeling (SEM) is used to test the hypothesis as well as the fitness of the whole model. SEM is a method that can be used to establish relationships among multiple variables. It has several advantages over simple correlation, such as considering the collinearity effect. It can also include any possible relationships among a set of variables. SEM is applied in the following procedures.

An initial model is specified and assessed by examining the whole model fit and individual parameter significance. Multiple criteria will be used to evaluate the whole model fitness (Hu and Bentler, 1999; Kaplan, 2000; Byrne, 2001), goodness of Fit Index (GFI) (Jöreskog & Sörbom, 1984), comparative fit index (CFI) (Bentler, 1990) and root mean square error of approximation (RMSEA) (Hu and Bentler, 1998; MacCallum and James, 2000). Rule of thumb recommended by scholars regarding the fit indexes is used to evaluate the model fit. Generally, GFI and CFI value above 0.9 are regarded as a good fit; RMSEA value less than 0.05 indicates good fit and value between 0.05-0.08 (Browne and Cudeck, 1993) represents reasonable fit. For normed Chi Square, Carmines and McIver (1981) recommended the value be below 3, but a value up to 5 also represents a reasonable fit (Wheaton et al., 1977; Marsh and Hocevar, 1985). If the model doesn't fit well, it should be re-specified. Those items whose path loading coefficients are insignificant ($\alpha > 0.05$) should be deleted for further test. In case all the measure coefficients are significant ($\alpha <= 0.05$), the item with smallest coefficient is deleted. The process should be one by one gradually. The process ends when the whole model satisfies all the fitness criteria and all individual measurement coefficients are significant. The evaluation criteria and standards

are summarized below:

- Coefficients for all paths are significant at 0.05 level
- X^2/df: <3 good fit, 3- 5 reasonable fit
- GFI and/or CFI: 0.9-0.95 good fit, > 0.95 superior fit
- RMSEA: <0.05 good fit, 0.05- 0.08 reasonable

RESULTS

The data analysis includes the test of four models. The first model (model-1) is based the conventional simple correlation. The second model (model-2) is based on multiple correlation with performance as dependent variable and four practices as independent variable. The third model (model-3) is based on the conceptual model in figure 2, i.e., all the hypotheses paths being included. The last model (model-4) will be the model deleting the no-significant paths gradually, if any. The testing results of the four models are summarized in table 5 and presented in details below.

Model 1 based on simple correlation

In model-1, each pair of the five variables are linked separately and simple bivariate correlation is calculated. The result is shown in table 5, the column of model-1. The result shows that all the correlation coefficients are significant. Based on the results from model-1, all the hypotheses should be accepted.

Table 5: The path significance (p) and model fitness tests of the four models

Hypotheses and paths			Model-1 Simple correlation		Model-2 Multiple correlation		Model-3 SEM (Initial)		Model-4 SEM (Final)	
H₁	Structural programs	→ Performance	0.00	✓	0.01	✓	0.44	✗	0.01	✓
H₂	Infrastructur al programs	→ Performance	0.00	✓	0.01	✓	0.92	✗	/	
H₃	Infrastructur al programs	→ Structural programs	0.00	✓	/		0.00	✓	0.00	✓
H₄	Technology	→ Structural programs	0.00	✓	/		0.04	✓	0.00	✓
H₅	Technology	→ Infrastructural programs	0.00	✓	/		0.00	✓	0.00	✓
H₆	Technology	→ Performance	0.00	✓	0.35	✗	0.64	✗	/	
H₇	Auto work org.	⬚ Technology	0.00	✓	/		0.00	✓	0.00	✓
H₈	Auto work org.	→ Structural programs	0.00	✓	/		0.89	✗	/	
H₉	Auto work org.	→ Performance	0.00	✓	0.71	✗	0.28	✗	/	
H₁₀	Auto work org.	→ Infrastructural programs	0.00	✓	/		0.00	✓	0.00	✓
SEM Model fitness indexes			n/a		X^2=858 X^2/df=3.15 CFI=0.95 RMSEA=0.088		X^2=516 X^2/df=1.94 CFI=0.98 RMSEA=0.06		X^2=518 X^2/df=1.92 CFI=0.98 RMSEA=0.057	
Model fitness test (Figure)			n/a n/a		Not (Cf., Fig.2)		Not (Cf., Fig.3)		Yes (Cf., Fig.4)	

Note: ✓: significant with p<0.05, ✗: not significant with p>0.05, /: the path was not specified or deleted due to insignificance

Model 2 based on multiple correlation

However, simple correlation does to take collinearity into consideration. This is proved by the test of model-2, which is based on multiple correlation. Model-2 is specified with performance as dependent variable and the four practice variables as independent simultaneously. The SEM model fitness test shows that only two paths are significant while two others are not significant as shown in table 5, the column of model-2. Different results can be observed in the two models. According to the SEM principle, as long as there is a non-significant path, the whole model does not fit well and no conclusion should be drawn. The reason is that the interrelations among the four practice variables have not been considered yet. This interrelationship may influence the relationship among practice and performance, as will be illustrated in the model-3 and 4.

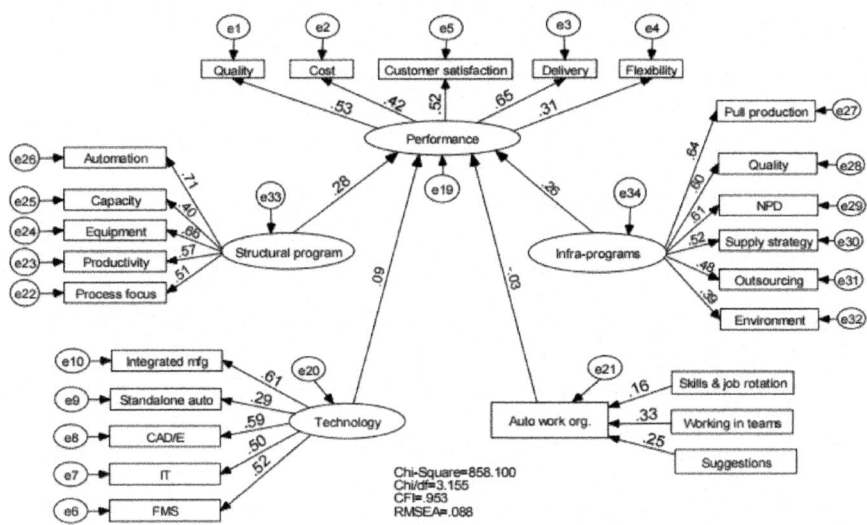

Figure 2: The test of model-2 based on the multiple correlation principle

Mode-3 & 4 based on SEM

Model-3 is specified based on the conceptual model (cf., figure 1) of manufacturing strategy and incorporates all the possible hypotheses among the five variables. It is the initial specified model for testing. The test result of model-3 is shown in the column of model-3 in table 5. The details are shown in figure 3. The test shows that five paths are not significant. Obvious differences can be found between model-2 and model-3. In model-2, the paths for H1 and H2 are significant but not significant in model-3. According to the SEM principle, as long as there are non-significant paths, the whole model does not fit well and no conclusion can be drawn.

In the next step, the non-significant paths are removed one by one GRADUALLY and the model is tested again. The principle for removing non-significant paths should follow the principle from the least non-significant to the next least non-significant each by each. The reason is that removing one of the paths may change the path significance of other remaining paths. In this case, the path for H2 (p=0.92) should be removed from the model first. Then the path for H8 (p=0.89) is removed. The process continues until all the remaining paths are significant and the whole model fits well. Finally a model-4 is obtained as shown in figure 4. In this model, all the paths are significant and the whole model passes the fitness test as well. Therefore, conclusion can be drawn from model-4.

According to the results from model-4, it can be found that among the 10 hypotheses, four hypotheses are rejected and six are accepted, as shown the column of model-4 in table 5 as well as figure 5. Hypotheses 1, 3, 4, 5, 7 and 10 are accepted, while hypotheses 2, 6, 8 and 9 are rejected.

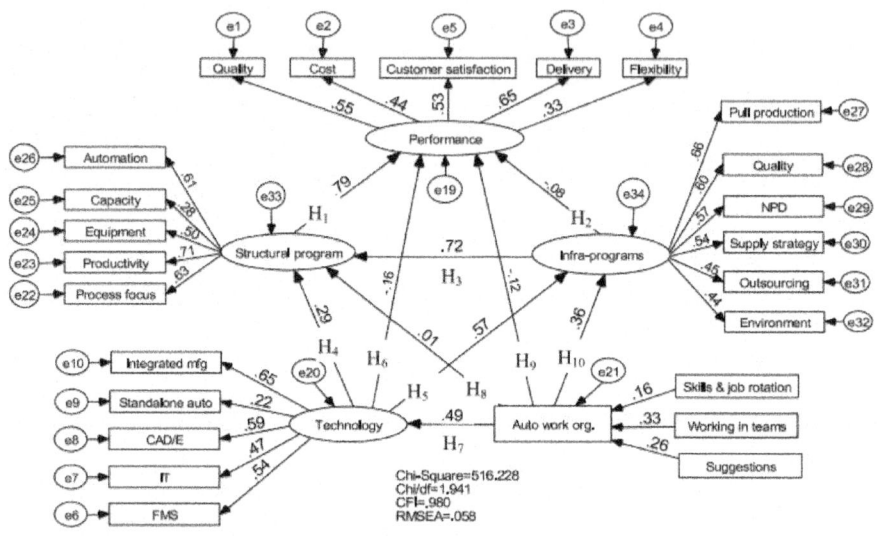

Figure 3: The specified model (model-3) and test result

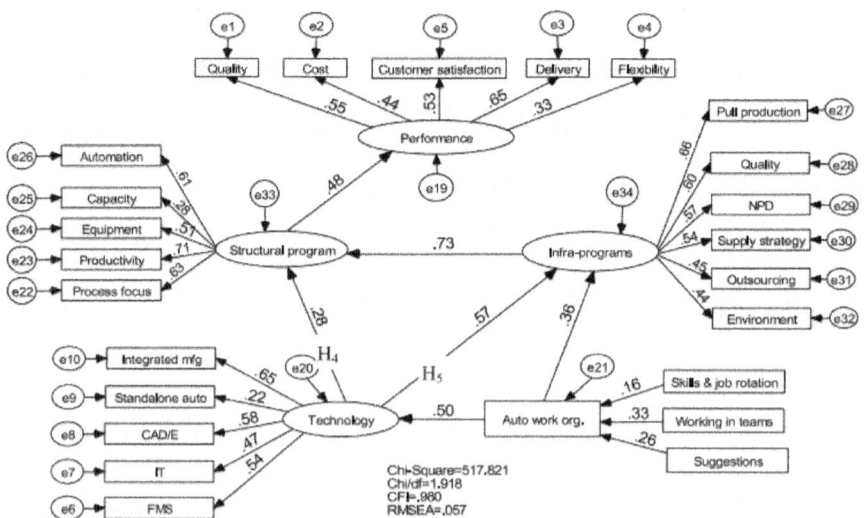

Figure 4: The modified model (model-4) by gradually deleting no-significant paths

DISCUSSIONS AND IMPLICATIONS

The research finds that structural programs are the only practice that directly contributes to manufacturing performance, while other three dimensions such as infrastructural programs, technology and organization contribute indirectly through structural programs. The research results trigger the following discussions.

Manufacturing process is the core

This research reveals that the improvement programs that are related to the physical process directly contribute to manufacturing performance. The structural programs work on the manufacturing process. Therefore, the process is the core and direct factor that explains manufacturing performance. This can be supported by another stream of structural research on quality management. The research based on the USMBQA framework is also a structural model and produced very valid and reliable research results. For most of the research based this model, process management is directly correlated with performance (Kaynak, 2003, Meyer and Collier 2001, Pannirselvam and Ferguson 2001, Wilson and Collier 2000). The implication is very clear. To improve the manufacturing performance, it is critical to improve the manufacturing process.

Infrastructure is the basis

It is surprising that infrastructural programs like quality management, full production etc do not directly contribute to manufacturing performance. This is opposite to many previous studies on the relationship between quality management and performance. However, if looking at the research models in previous research, it will be possible to explain the difference. In previous research, only part of the programs is investigated and other relevant factors such as structural programs are ignored. When simple correlation or multiple correlation is used in this research, the infrastructural programs are found to be positively correlated with performance (cf., model-1 in table 5). Then the conclusion will be different.

The explanation is that infrastructural programs are useful. However, they do not contribute directly to performance but through the structural programs. The path loading between infrastructural to structural programs is very high (0.73) and very significant (p=0.001). These infrastructural programs are for the establishment of infrastructure. They support the manufacturing structural technical process. The finding implies that whatever infrastructural programs to be implemented, the evaluation may not be whether it directly contributes to performance, but the requirement of the process or programs related to the structural side of the process.

Technology and organization are useful, but not directly contributing

Technology is not found to be directly correlated with performance. In the past 20 years, AMT has been widely used by manufacturing companies all-over the world. However, world-wide research found that not all AMT perform as expected. Some AMTs performs "satisfactory", but did not produce the full benefits. Other AMTs perform well on the shop floor level, while the business performances of the companies were not improved (Voss, 1988). All these problems have caught the attention of both researchers and practitioners.

Since the beginning of the 1980s, management of technology, especially implementation of AMT, has been a hot topic (Gerwin, 1982; Voss, 1988). The relationship between AMT and performance was investigated conceptually (Macbeth, 1989, p.71; Bishop and Schofield, 1989, p.44), by case studies (Sohal, 1996; Sun, Hjulstad and Frick, 1997) and by survey (Sun 2000, Small, 1998). Recent empirical research does not found that the use of AMT has direct impact on business or manufacturing performance (Swamidass and Kotha, 1998). The research by Beaumont et al (2002) intents to investigate AMT investment and performance in foreign-owned and Australian domestic companies. They did not conclude whether the AMT is significant related to performance. Sun (2000) found that little linear relationship exists between AMT and performance. The result from this research provides a reasonable explanation. Future research is needed to investigate the detail relationship between AMT and structural infrastructural action program. For example IT and supply chain management is one of the topic recently attracts researchers' attention.

Methodological implications

In this research, four different models are tested for the same set of hypotheses tests. Obvious differences are found among the four models. The differences have significant implications for selection of research methods on relationships among multiple variables. Simple correlation is simple and visual. However, its main limitation is the ignorance of the collinearity effects among variables. It can be used for identity or specify the preliminary model or explorative research at preliminary stage. Multiple correlation has the advantage of taking collinearity into consideration. However, it does not cover the interactions among the independent variables. If there are such interactions, multiple correlation results may not be reliable. Structural Equation Modeling (SEM) method is a good method since it covers collinearity effects and interactions among all the variables. As a result, it is more reliable for investigating relationships among multiple variables. More research on operations management, technology management and quality management are using more SEM to investigate multi-variant relationships (Kaynak, 2003, Meyer and Collier 2001, Pannirselvam and Ferguson 2001, Wilson and Collier 2000).

CONCLUSIONS, LIMITATIONS AND FUTURE RE-SEARCH

The research in this paper has investigated the complex relationship among manufacturing practice and manufacturing performance. It is based on a structural model that incorporates all the possible linkages among practices and performance. The research may have the following contribution to the literature on practice-performance linkage. First, the research is based on the conceptual framework of manufacturing strategy, therefore, the model prevents from ignoring any possible linkages. Second, the data analysis is conducted with all available methods so that differences and limitations of simple and multiple correlation analysis are identified. Finally, the research produces several different results which are worthwhile to be considered in research in operations management.

The main message from this research is that not all practices may directly contribute to performance. It is the structural programmes that directly contribute to performance. Whatever other programs or technologies or organizational practices to be implemented, the final goal is to improve the manufacturing process. If the process is not improved, the contribution of other practices may not be realized.

Since the research aims to be comprehensive and holistic, the scope of the paper is pretty wide. The ten hypotheses may not be fully discussed conceptually. The implications are not fully explored for each sub-relationship. Page and words limitation may also contribute to this weakness. However, in future research which looks at a sub-relationship, for example, between technology and structural programs, the conceptual part should be enhanced.

Some of the sub-relationships have been well studied. For example, the relationship between technology and HRS/organization has been studied insensitively in the past decades. However, future research may include the following topics, the relationship between technology and structural programmes, the relationship between technology and infrastructural programs, as well as the relationship between structural and infrastructural programs.

The research provides a conceptual model and data analysis approach for investigating practice-performance relationships. Triangulation research based on the model is welcomed and appreciated to cross-proof the validity of the research method. Based on this method, a series of comparative studies can be conducted, for example, between mass and job-shop process, between Small and Media Enterprises (SME) and larger companies, and between developed and developing countries.

REFERENCES

1. Ayres, Robert (1991) Computer Integrated Manufacturing Volume I: Revolution in Progress, CHAPMAN & HALL, London.

2. Beaumont, N., Schroder, R. and Sohal, A. (2002) Do foreign-owned firms manage advanced manufacturing technology better? International Journal of Operations & Production Management;Vol. 22, No. 7/8, pg. 759-772.

3. Beekun, R. I. (1990) "Assessing the effectiveness of socio-technical interventions: Anitdore or fad?", Human Relations, Vol. 42, pp.887-897.

4. Beekun, R. I. (1990) "Assessing the effectiveness of socio-technical interventions: Anitdore or fad?", Human Relations, Vol. 42, pp.887-897.

5. Bessant, J. and Haywood, B., (1988), "Islands, archipelagos and continents: progress on the road to computer-integrated manufacturing", Research Policy, Vol. 17, 349-362.

6. Byrne, B.M., 2001. Structural Equation modeling with AMOS: Basic Concepts, Applications and Programming. Lawrence Erlbaum Associates, Publishers, NJ.

7. Chase, R. B. and Aquilano, N. (1997) Production and Operations Management, IRWIN, Chicago.

8. Chase, R. B., Aquilano, N. and Jacobs, R. (2001) Operations management for competitive advantage (9th ed.) Boston : McGraw-Hill/Irwin

9. Cohen, J. (1988), *Statistical power analysis for the behavioral sciences*, 2nd ed. Hillsdale, N.J. : L. Erlbaum Associates, 1988.

10. Cua, K. O. McKone, K. E. and Schroeder, R. G. (2001) Relationships between implementation of TQM, JIT, and TPM and manufacturing performance. Journal of Operations Management. Vol.19, No. 6, pp. 675-694.

11. Dasa, A. and R., Narasimhan (2001) Process-technology fit and its implications for manufacturing performance, Journal of Operations

Management 19 (2001) 521–540 de Vaus, D. A., (1993), *Survey in Social Research* (3rd ed.), UCL PRESS.

12. Fix-sterz, Jutta, Gunter Lay, Rainer S., Jurgen W. (1987), *Flexible Manufacturing Systems and Cells in the Scope of New Production Systems in Germany,* FAST Occasional paper No.135.

13. Flynn, B. B., Schroeder, R. G., and Sakakibala, S. (1994), "A framework for quality management research and an associated measurement instrument", *Journal of Operations Management,* Vol. 11, pp.1339-1366.

14. Gerwin, D. and H. Kolodny (1992) *Management of advanced manufacturing technology: strategy, organisation, and innovation,* Wiley & Sons, N.Y.

15. Goetsch, D. L., (1990) *Advanced Manufacturing Technology,* Delmar Publisher Inc., New York Greenwood, N. R. (1988) I*mplementing Flexible Manufacturing System,* Macmillan Education, London.

16. Groover, M. P. (1987), *Automation, Production Systems, and Computer Integrated Manufacturing,* Prentice-Hall, Inc., New Jersey.

17. Hanson, P. and Voss, C.A., (1993), Made in Britain, the true state of Britain's manufacturing industry, IBM Ltd/London Business School, Warwick, UK.

18. Hayes, R. H., Wheelwright, S. C. and Clark, K. B., (1988), Dynamic manufacturing, creating the learning organization. The Free Press, New York, NY.

19. Hayes, R.H. and Wheelwright, S.C. (1984) Restoring Our Competitive Edge. Wiley, New York. Hayes, R.H. and Wheelwright, S.C., (1984), Restoring our competitive edge, Wiley, New York. Haywood, B. (1990) "CIM: technologies, organisations, and people in transition", *Proceedings of the final IIASA conference on CIM,* Luxembourg, Austria.

20. Hill, T., (1995), Manufacturing strategy: Text and cases, Macmillan Press, London.

21. Hu, L., Bentler, P.M., 1998. Fit indices in covariance structure modeling: Sensitivity to underparameterized model misspecification. Psychological Methods 3, 424-453.

22. Hu, L., Bentler, P.M., 1999. Cutoff criteria for fit indexes in covariance structure analysis: conventional criteria versus new alternatives. Structural Equation Modeling 6(1), 1-55.

23. Jöreskog, K.G. and Sörbom, D., 1984. LISREL-VI user's guide (3rd ed.). Scientific Software, Mooresville, IN.

24. Kaynak, H. 2003. The relationship between total quality management practices and their effects on firm performance. Journal of Operations Management 21, 405–435.

25. Kim J.S., Arnold, P., 1996. Operationalizaing manufacturing strategy: an exploratory study of constructs and linkage. International Journal of Operations & Production Management 16(12), 45-73.

26. Kopelman, R. E. (1986) *Managing Productivity in Organizations, McGraw, NY.*

27. Kotha, S and P M Swamidass (2000) Strategy, advanced manufacturing technology and performance: Empirical evidence from U.S. manufacturing firms, Journal of Operations Management. Vol. 18, No. 3: 257

28. Kotha, S. and P. M., Swamidass, (1998), "Advanced Manufacturing Technology uses: exploring the effect of the nationality variable", *International Journal of Production Research*, Vol. 11, pp.3135-3146.

29. Lee, R. J. V. and Leonard, R. (1990), "Changing role of humans within an integrated automated guided vehicle system", *Computer-Integrated manufacturing Systems,* Vol. 3. No.2, pp.115-120.

30. Leong, G.K. and Ward, P.T., (1995), The six Ps of manufacturing strategy, International Journal of Operations & Production Management, Vol. 15, No. 12, p 32-45.

31. Lewis, M.A. and Roehrich, J. (2009), ``Contracts, relationships and integration: towards a model of the procurement of complex performance", *International Journal of Procurement Management,* Vol. 2 No. 2, pp. 125-142.

32. Lindberg, P. (1990) *Manufacturing Strategy and Implementation of Advanced Manufacturing technology,* Ph.D. Dissertation, Chalmers University of Technology, Gothenburg, Sweden.

33. Lindberg, P., Voss, C.A., Blackmon, K.L. (Eds), 1998. International Manufacturing Strategies: Context, Content and Change. Kluwer Academic Publisher, Boston.

34. MacCallum, R., 1986. Specification searches in covariance structure modeling. Psychological Bulletin 100(1), 107-120.

35. MacCallum, R.C., James, T. A., 2000. Applications of structural equation modeling in psychological research. Annual Review of Psychology 51, 201.

36. Marsh, H.W. and Hocevar, D., 1985. Application of confirmatory factor analysis to the study of self-concept: first- and higher-order factor models and their invariance across groups. Psychological Bulletin 97, 562-582.

37. McDermott, C.M., Stock, G.N. (1999) Organizational culture and advanced manufacturing technology implementation. Journal of Operations Management 17 (5), 521–533.

38. McDermott, C.M., Stock, G.N., 1999. Organizational culture and advanced manufacturing technology implementation. Journal of Operations Management 17 (5), 521–533.

39. Meyer, S. M. and A. Collier. 2001. An empirical test of the causal relationships in the Baldrige Health Care Pilot Criteria. Journal of Operations Management 19(4), 403:425.

40. Ng, K.C. and Hung, I. W., (2001), A model for global manufacturing excellence, Work Study, Vol. 50, No. 2, p 63-68.

41. Noori, H. (1990) *Managing the Dynamics of New Technology, issues in manufacturing management,* Prentice Hall, Englewood Cliffs, N.J.

42. Nunnally, J. C. (1978), *Psychometric Theory*, McGraw-Hill Publishing Company, New Yor. Nunnally, J.C., Bernstein I.H., 1994. Psychometric theory. McGraw-Hill, New York, 510-512. Pannirselvam, G.P., Ferguson, L.A., 2001. A study of the relationships between the Baldrige

43. categories. International Journal of Quality and Reliability Management 18 (1), 14–34. Parthasarthy, R. and S. P. Sethi (1992) "The Impact of flexible automation on business strategy and organizational structure", *Academy of Management Review,* Vol. 17,No:1. pp.86-112.

44. Peterson, R.A., 1994. A meta-analysis of Cronbach's coefficient alpha. Journal of Consumer Research 21(2), 381-391.

45. Riis, O.J. (1992) "Integration and Manufacturing Strategy". Computer in Industry, vol. 19, 37-50.

46. Samson, D., Sohal, A. and Ramsay, E., (1993) "Human resources issues in manufacturing improvement initiatives: case study experiences in Australia", *The International Journal of Human Factors in Manufacturing,* Vol.3, No. 2, pp.153-152.

47. Saraph, J. V., and Sebastian, R. J. (1992), "Human resources strategies for effective introduction of advanced manufacturing technologies (AMT)", *Production and Inventory Management Journal,* Vol.33, pp.764-770.

48. Saraph, J. V., Benson, P. G., and Schoeder, R. G. (1989), "An instrument for measuring the critical factors of quality management", *Decision Sciences*, Vol. 20, pp.810-829.

49. Schonberger, R. J., (1982), Japanese Manufacturing Techniques, Nine Hidden Lessons in Simplicity, The Free Press, New York NY.

50. Schonberger, R.J., (1986), World Class Manufacturing: The Lessons of Simplicity Applied, The Free Press, New York, NY.

51. Schroeder, R.G., Anderson, J.C., and Cleveland, G., (1986), The content of manufacturing strategy: An empirical study, Journal of Operations Management, Vol. 6, No. 4, p 405-415

52. Singh, N. (1996) *System Approaches to Computer Integrated Design and Manufacturing,* John Wiley & Sons, Inc., N.Y.

53. Skinner, W., 1969. Manufacturing-missing link in corporate strategy, Harvard Business Review 47(3), 136-145.

54. Sobek II, D. K., Liker, J. K. and Ward, A. C. 1998, Another look at how Toyota integrate product development, *Harvard Business Review*, July-August issue, pp.69-78.

55. Sun, H. (2000) "Current and Future Patterns of Using Advanced Manufacturing Technologies", *Technovation, The International Journal of Technological Innovation and Entrepreneurship,* Vol.20, No.11, pp.631-641.

56. Sun, H. (2001) "Human Resources Development in Integrated Manufacturing Systems",

57. Integrated Manufacturing System, Vol. 12, No.3, pp.195-204

58. Sun, H. and Cui, H. (2002) The alignment between manufacturing and business strategies: its influence on business performance. Technovation 22, 699-705.

59. Sundstrom, E., DeMeuse, K.P., and Futell, D. (1990), "Work teams", *American Psychologist*, Vol. 45, pp. 120-133.

60. Swamidass, P.M. and Newell, W.T. (1987) "Manufacturing strategy, environmental uncertainty and performance": a path analytic model. Management Science, vol.33, 509-524.

61. Swamidass, P.M., Kotha, S. (1998) Explaining manufacturing technology use, firm size and performance using a multidimensional view of technology. Journal of Operations Management 17 (1), 23–37.

62. Tseng, H.C., Ip, W.H., and Ng, K.C., (1999), A model for an integrated manufacturing system implementation in China: a case study, Journal of Engineering and Technology Management, 16, p83-101.

63. Tunalv, C. (1991) Manufacturing strategy in Sweden engineering industry", Ph.D. thesis, Chalmes University of Technology, Sweden.

64. Wheaton, B., Muthen, B., Alwin, D.F. and Summers, G.F., 1977. Assessing reliability and stability in panel model. In: Heise, D.R. (Ed), Sociological Methodology. Jossey-Bsaa, San Francisco, 84-136.

65. Wheelwright, S. C., (1984), Manufacturing strategy: Defining the missing link, Strategic management Journal, 1, p 77-91.

66. Wilson, D.D., Collier, D.A., 2000. An empirical investigation of the Malcolm Baldrige National Quality Award causal model. Decision Sciences 31 (2), 361–390.

67. Woodward, Joan (1965) *Industrial Organization: Theory and practice.* London: Oxford University press.

Chapter 11

ADVANCED MANUFACTURING TECHNOLOGY PROJECTS JUSTIFICATION

Josef Hynek and Václav Janeček

University of Hradec Králové, Faculty of Informatics and Management Czech Republic

INTRODUCTION

Manufacturing companies worldwide are pressurized to undergo a transformation processes in order to compete more effectively and under these circumstances advanced manufacturing technology (AMT) is considered to be a very important tool improving their ability to succeed with their products on extremely competitive international markets. It is widely believed that AMT has a great potential to provide the respective companies by a whole variety of tangible as well as intangible benefits and the reduction of production cost, increased volume of production, improved quality as well as better safety at work are usually amongst the most mentioned ones. On the other hand it is also generally understood that the adoption of AMT requires a high level of initial investment and also the level of risk associated with the implementation of the AMT project is higher especially when the particular company lacks relevant experience. Moreover the payback period of advanced manufacturing technology investment is as a rule longer than the payback period of rather traditional and usually less expensive technology. That is why the process of adoption and utilization of advanced manufacturing technology has been carefully studied and examined in last two decades and numerous studies were published in order to provide some guidelines for managers of manufacturing companies with the view of helping them to make good and well-founded decisions. We also strongly believe that it is important to study the respective processes when the crucial decisions about AMT projects justification resulting into their practical implementation or on the contrary their rejection are made. The deep comprehension of the fundamentals of these processes allows us to derive the appropriate pieces of knowledge that could turn out to be helpful to technology specialists. We will present selected results of two extensive surveys targeted on adoption and utilization of advanced manufacturing technology that were carried out recently in the Czech Republic in this chapter. We will focus on the phase of advanced manufacturing technology project economic justification and findings ascertained in the Czech Republic will be compared with the outcomes of analogous surveys that were carried out earlier in the United Kingdom and the United States of America. We will demonstrate there are many problems of advanced manufacturing technology projects justification that we have in common in all the above mentioned countries and we believe that technology specialists as well as managers worldwide could learn from issues presented and discussed here. Based on our results we suppose that technology specialists empowered in advance by broader insight of what kind of difficulties to anticipate they should be able to prepare their AMT projects accordingly and to improve their chance to get the management approval for the project financing and its implementation.

PROBLEM DEFINITION

We have already pointed out that advanced manufacturing technology is rather expensive and the relevant project is associated with a higher degree of risk. Therefore the proper and sound justification of the investment decision is required. If the project is incorrectly undervalued and it does not get through the justification process, the company will miss the opportunity to derive potential benefits and its competitiveness might be jeopardized. On contrary, if the project is overvalued because of technology enthusiasm or because of the other reasons, it will be implemented and then it is likely that it will not meet the initial expectations. It will cause a disappointment and furthermore, it will complicate the justification process for further AMT projects that will be perceived through biased lens as the former experience was not a positive one. Whatever the motives are, we can see that the both problems, underestimation as well as overestimation of AMT projects, are terribly wrong and unfortunate. That is why the appropriate methods used for AMT projects justification and their proper utilization are extremely important. It is widely accepted that there are three general groups of investment appraisal techniques - economic approach, analytic approach and the strategic approach. The economic justification approach seems to be very natural and straightforward one and perhaps that is why it is so wide-spread in relevant companies worldwide. AMT investment has to be financially sound and viable because such a project competes for limited resources with many other projects. Therefore various financial and accounting justification techniques such as payback period (PP), return on investment (ROI), net present value (NPV), and internal rate of return (IRR) are frequently used by managers in order to assess the economic aspects of the project. However, many researchers argue that these methods support decisions that are sensible when viewed in isolation and they do not always indicate the best action when we take into account the whole organizational context (Chan et al., 2001). Furthermore, these methods could be misleading when employing too short payback periods or too high discount rates, neglecting various benefits of the new AMT system or being unable to quantify them properly in financial terms. To overcome the problems inherent in using purely economic appraisal techniques, analytic and strategic appraisal approaches have been promoted. The analytic justification approaches are predominantly quantitative but more complex than the economic techniques. It is believed that especially when intangible benefits are taken into account, these techniques can be far more appropriate by being more realistic, offering better reflection of reality and taking more factors into consideration (Meredith & Suresh, 1986). Various scoring and ranking models could be used including some traditional optimization techniques as well as risk analysis approaches. It is clear that the transformation process from the decision problem to the particular model involves a great deal of simplification and many important factors could be easily overlooked. Furthermore, models involving various weights of individual factors are rather vulnerable to bias brought along with subjective judgments. The strategic justification approaches tend to be less technical that economic and analytic methods, but it should be stressed that they are quite often used in combination with them. The main advantage of the strategic approaches is their direct linkage to the goals of the company. Criteria such as meeting the business objectives, comparison with competitors, the retention or attainment of competitive advantage and industry leadership might be utilized as suitable factors for the relevant decision making processes where AMT projects are scrutinized. Of course, it would be unwise to assign too much importance to strategic justification methods and to overlook the economic and tactical impact of the project. That is why recent studies have promoted hybrid approaches based on suitable combination of economic, analytic and strategic appraisal techniques (see (Raafat, 2002)). We will focus on the economic justification techniques in the rest of this chapter. These techniques seem to be widely used in manufacturing companies worldwide when the decision concerning AMT investments should be made. It is quite natural because the cost of such project is usually well known (although it could be very easily underestimated too) and it is necessary to cover the cost by relevant revenues and various benefits. We will show some typical problems related to the utilization of economic justification techniques and we would like to stress that some researchers have even claimed that these techniques are inappropriate for evaluating AMT projects (Bucher & Lee, 2000).

LITERATURE REVIEW

There are many interesting papers describing various issues of AMT projects justification from different points of view. Perhaps the easiest way to get quickly oriented in the field is to start with a comprehensive bibliography on justification of AMT (Raafat, 2002) that cites over two hundred articles from a variety of published sources. Chan et al. (2001) concisely reviewed various approaches used in the process of investment appraisal of AMT and concluded that improved approach that would integrate the currently used evaluation approaches was needed. Abdel-Kader & Dugdale (1998) wrote an interesting paper reporting the results of a survey investigation into the investment decision making practices of large UK companies and their study focused especially on investments in AMT. On the other hand, Ariss, Raghunathan & Kunnathar (2000) published their findings concerned factors affecting the adoption of AMT in small manufacturing firms in the United States. Hofmann & Orr (2005) presented the results of their postal survey that was conducted amongst German manufactures and one part of their questionnaire was devoted to the assessment of AMT proposal too. Finally, we have decided to put forward the paper written by Small (2006) that summarizes the results of investigation on the justification of investments in AMT at US manufacturing plants. We proudly acknowledge that the biggest motivation to start our own investigations in the field of AMT in the Czech Republic came from the work of Lefley & Wharton (1993), Lefley (1994), and Lefley & Sarkis (1997). These authors examined carefully the investment appraisal processes in the United Kingdom and the United States of America. They carried out extensive surveys both in the UK and the USA in order to learn more about current practices in respect of capital investment in AMT projects, to identify if there were perceived difficulties in appraising these projects and to elicit the opinions of senior executives on the various issues related to AMT projects evaluation. Among other things they found out that AMT projects were evaluated by the simplest financial criteria that seem to be unsuitable in this respect. Moreover, they realized that financial directors do have many difficulties when assessing various benefits of AMT projects, and finally, that investment into AMT could be easily influenced by business culture where managers are under pressure to produce shortterm results. The first study in this field in the Czech Republic (Lefley et al., 2004) revealed that despite of many differences ascertained especially in the extent as well as the level of evaluated and implemented technology, where Czech manufacturing companies lagged behind their western competitors, there were many problems that were common for managers from all the three surveyed countries. These results fostered our interest to conduct the second survey in the Czech Republic in 2005 in order to identify the relevant changes in the results that were expected due to the quickly transforming Czech economy and its openness. And finally, we undertook the last survey in the Czech Republic in 2008 and we were interested in evaluation of AMT benefits this time. The results of this investigation are being carefully analyzed, processed statistically and we plan that we will be able to publish them later this year. However, the survey results described below have been derived from the first and the second survey only.

Survey Methodology

To keep in line with the earlier UK and US surveys which were used as a basis for comparison we have decided to employ the same questionnaire as Lefley & Wharton (1993) utilized earlier for their investigations. We translated their original English questionnaire into Czech language and verified its localization by means of a pilot survey. The original questionnaire comprised of three sections. Questions in the first part were intended to establish the level of implementation of AMT that had been achieved to date. Three levels of AMT were identified which correspond to the levels of sophistication proposed by Dornan (1987) and Meredith & Suresh (1986). Level 1 systems cover standalone projects e.g. robots, NC machines, CAD etc. Level 2 systems are linked systems e.g. linking together of a number of CNC machines, CAD/ CAM etc., and Level 3 systems are fully integrated systems including computer integrated manufacturing (CIM) and flexible manufacturing systems (FMS). In part number two of the survey the respondents were asked which techniques and criteria were used in capital project appraisal and what methods, if any, were used to measure and take into account project risk. Information was obtained about the measures used to assess the performance of senior executives as it appears that management in general is reluctant to make long-term risky investments (such as those in AMT) and prefers to invest in shortterm projects that show early profits and low risk (Lefley, 1994). The third part of the survey was designed to explore opinions about the need for AMT investment, the efficacy of the investment criteria used and the extent to which other factors and considerations had a bearing on capital investment decisions. We added one more additional section to the questionnaire that was used in the Czech Republic in 2005. It was devoted to the utilization of EVA (economic value added) indicator in our companies as there were some suggestions that there might be a relationship between utilization of this concept and investment behavior of manufacturing companies.

To assure a straightforward comparison of collected data in different countries we carefully followed the methodology used by our predecessors. The survey was aimed at those companies who, it was believed, would have had some experience in the appraisal of AMT projects and that the person who was asked to complete the questionnaire should have had a significant contribution to make in final investment decision. A number of databases were reviewed (with the main stress on data acquired from EDB and Czech business register) to identify the largest manufacturing companies. As we wanted to restrict the survey to 'large' Czech manufacturing organizations, we finally chose sample size of 416 firms in 1999. Within our last survey we have decided to include also the middle sized Czech manufacturing firms and so we have increased the sample to 1030 in 2005. Our first postal survey started at the end of 1998 and of the 416 questionnaires sent out 92 was returned giving a response rate of 22.12%. A usable sample of 79 completed questionnaires with a response rate of 19.0% was considered to be reasonable under the existing circumstances. The second postal survey has been conducted from January till April 2005 and 1030 questionnaires were sent out and 135 have returned, 3 of them were unusable. We can see that the rate of response is 12.8% only which is significantly lower rate that the one we achieved in 1999. The reason that we did not reach comparable numbers with our former survey could be explained by the fact that in our current survey the middle sized firms were addressed too. This article deals with the selected results derived from the first three parts of our questionnaire only and due to limited space we cannot dwell on the other issues here. Readers who are interested in further details are advised to look at (Hynek & Janeček, 2007) or (Hynek & Janeček, 2008).

Survey Results

and discussion The main part of this section will be devoted to economic justification of AMT projects, but we believe that the facts we will present here should be perceived in a broader context. That is why we will outline basic facts concerning the experience of Czech companies in the area of AMT projects evaluation as well as the levels of implemented technology that were achieved by surveyed companies. Furthermore, selected personal opinions of managers will be put forward in order to show some important problems and difficulties that could significantly influence the chance of AMT projects to pass successfully through the evaluation process.

Appraisal Experience and Level of AMT

First of all, from the point of view of further discussions concerning AMT projects justification it could be worthwhile to learn more about the experience of Czech manufacturing companies in the area of AMT projects evaluation. We can see from table 1 that 82.3 % in year 1999 and 78.3 % in year 2005 of Czech manufacturing companies claimed they had evaluated AMT projects.

These numbers are significantly lower than results described by Lefley and Sarkis (1997) who reported that 99.3 % of UK and 96.7 of US companies stated that they had evaluated AMT projects over the past ten years. It is clear that Czech managers are less experienced in this respect. Moreover, we have to take into account the time difference among the surveys.

Furthermore, 84.8 % of Czech manufacturing companies in year 1999 and 92.3 % in year 2005 stated that they expect to consider such projects within the next ten years. Once again, comparing these findings with 97.1 % of respondents in UK and 99.2 % in US (Lefley & Sarkis, 1997), there is a significant difference here despite the fact that the latter result ascertained in the Czech Republic might be considered as a positive signal evidencing the raising awareness of the importance of AMT projects amongst Czech managers.

Table 1: Companies that had evaluated AMT investment proposals

AMT projects evaluated	1999	2005
Number of companies	65	101
Percentage	82.3 %	78.3 %
Total number of companies	79	132

Secondly, we were interested in the level of manufacturing technology that was taken into consideration and consequently the level of technology that was actually implemented in the surveyed companies. There were some thoughts that massive foreign investment into transforming and quickly developing Czech economy during last two decades could accelerate the processes of adoption of advanced technology in manufacturing companies. The respective results are summarized in the table number 2 and 3 below.

Table 2: Level of evaluation of AMT projects

% number of companies that evaluated AMT project at:	1999 [%]	2005 [%]
Level 1 (stand alone projects)	57.0	40.4
Level 2 (linked systems)	35.4	41.3
Level 3 (fully integrated systems)	15.2	18.3

Table 3: Level of implementation of AMT projects

% number of companies that implemented AMT project at:	1999 [%]	2005 [%]
Level 1 (stand alone projects)	51.6	45.0
Level 2 (linked systems)	33.9	36.9
Level 3 (fully integrated systems)	14.5	18.0

It is clear that many projects that were originally planned on a higher level were unable to reach the stage of practical implementation and only the restricted version of the project (on a less sophisticated level of technology) was carried out. There is an obvious positive tendency that we can see in the table number 2 as the percentage of Czech manufacturing companies that evaluated the higher level AMT project proposals have been increased in 2005. The same is true for the implementation stage but comparing these results internationally we have to admit significant differences in respect of stages reached by UK, US and Czech manufacturing companies in relation to the evaluation and implementation of AMT projects. For example, taking into account the results of British and US surveys (Lefley & Sarkis, 1997) it is unmistakable that significantly greater number of UK (55.1%) and US companies (50.9 %) had evaluated the most sophisticated projects (on the third level) while the Czech companies have in majority only the first and the second level experience (we can see from table 2 that only 18.3 % of companies reached the third level technology evaluation experience).

Moreover, based on the results shown in table number 3 it is evident that Czech manufacturing companies are lagging behind their British and American competitors in the adoption of advanced manufacturing technology. The contrast is especially visible when focusing on the most advanced fully integrated systems (only 18.0 % of Czech firms implemented them comparing to the 43.0 % in the UK and 43.4 % in the USA). Moreover, as we can see in table number 3 the situation in the Czech Republic has not changed very much between 1999 and 2005 and therefore the gap is still huge (Hynek & Janeček, 2007). It is obvious that the high level of foreign direct investment in the Czech Republic did not fetch along anticipated acceleration of advanced technology adoption in manufacturing companies and the achieved levels of AMT implementations are lower than those previously observed in the UK and USA. Unfortunately, as we will discuss below, we have found that reasons for this unfavorable position of Czech manufacturing companies does not lie with lack of investment money only but it might be deeply rooted in management attitudes too.

AMT Projects Justification

Our findings that were described in the previous section clearly demonstrated that the level of AMT evaluation as well as its utilization in the Czech Republic is lower than the levels observed earlier in the UK and the USA. Furthermore, we have indicated that the process of AMT adoption might be influenced by management attitudes towards technology investment in general and, of course, the particular evaluation and justification approaches chosen by the relevant decision makers could be seen as a direct and straightforward way of influencing the outcome of the AMT projects evaluation processes. Some researchers and technology promoters expressed their concerns over conventional appraisal techniques such as payback, return on investment, or net present value, claiming that these techniques are inadequate and biased against technological investment in general (see, for example, Chan et al, 1999). Their criticism is based on assumption that while the cost of the proposed AMT project is in general easily quantifiable, there are many benefits that are very often difficult to estimate. Moreover, as AMT projects tend to be of long-term nature and sometimes even full deployment of particular AMT project requires substantial time period, the profits cannot be expected in short time and that is why the decisions on these projects require a long-term perspective. Subsequently, if the chosen appraisal method is well known for favoring short term profits, the relevant investment decision that is based on such method is easily predictable. Table 4 shows financial criteria used to assess AMT projects by financial directors of Czech manufacturing companies. It is obvious that more than 60 % of Czech managers employ the simple non-discounted cash flow payback period (non-DCF Payback) as the criterion to decide whether to finance such a project or not (see table 4 for more details) and more than 70 % of them use discounted version of payback (DCF Payback). And it is exactly payback criterion that is often criticized and attacked for its inappropriateness regarding AMT projects. Naturally, this criterion prioritizes projects capable of early repayment of initial expenses while as a rule capital intensive AMT projects tend to be slow in generating positive net cash flows. Indeed, many argue that the utilization of the payback method virtually guarantees the rejection of projects such as AMT (Lefley et al, 2004). On the other hand it has to be stressed, that the problem is not caused by the criterion itself, but it arises when a short payback period is requested by the company management. As we can see in (DeRuntz and Turner, 2003), while the western companies generally accept a payback period of 1 to 5 years as a reasonable amount of time to recover the initial cost, the Japanese companies are much more flexible in this respect as they use the payback method more as a performance measure than as a rigid financial criterion that must be met.

Table 4: Financial appraisal criteria

Financial appraisal criteria used	1999 [%]	2005 [%]
IRR/yield	31.1	35.5
NPV	45.9	38.7
DCF Payback	71.6	76.6
Other DCF	5.4	10.5
Non-DCF Payback	63.5	62.1
ARR	35.1	23.4
Other non-DCF	1.4	2.4

It has been anticipated that many companies would use more than one criterion and that is why we have made inquiries regarding the number of financial appraisal criteria being used and their importance. The results are summarized in tables 5 and 6. It should be noted that percentages given in table 6 below add up to more than 100 % because some respondents gave equal first ranking to more than one technique.

Table 5: Number of different financial appraisal methods used

Number of methods used	1999 [%]	2005 [%]
1	23.0	22.6
2	32.4	33.1
3	20.3	25.0
4 or more	24.3	19.3

Table 6: Percentage of companies ranking criteria first

Criteria ranked first or first equal	1999 [%]	2005 [%]
IRR/yield	5.4	9.7
NPV	28.4	17.7
DCF Payback	51.4	58.1
Other DCF	1.8	4.0
Non-DCF Payback	43.2	62.1
ARR	13.5	8.9
Other non-DCF	0.0	1.6

It is definitely a positive ascertainment that more than 40 % of financial directors use more than two financial criteria when assessing an AMT project proposal. On the other hand it is evident that every fifth company relies on single criterion only and here of course the important issue is which criterion is employed in these cases. We can see from table 6 that Czech managers without any doubt prefer both versions of payback criterion. Moreover, within our last survey the above mentioned and criticized simple non-discounted cash flow payback period (non-DCF PB) has been ranked as the most important one in the Czech Republic (62.1 % in 2005) while discounted version of this criterion came second (58.1 %). It should be emphasized that there is a huge gap afterwards as the third most important criterion (net present value) maintained its position from 1999 but it was ranked as the most important criterion by 17.7 % of Czech managers only in 2005. Comparing these results with earlier ascertainments of Lefley and Sarkis (1997) we could find out that non-discounted cash flow payback period (non-DCF PB) was ranked as the number one criterion in the United Kingdom (38.5 %). American managers inclined to use more sophisticated methods that make allowance for the time value of money and that is why DCF Payback (ranked first by 33.3 % of managers) was closely followed by internal rate of return (IRR) that was preferred by 28.2 % of US managers. From this point of view it is quite interesting that IRR is rather popular amongst British managers too (28.0 %), while only 5.4 % of Czech managers in 1999 and 9.7 % in 2005 marked it as the most important criterion. It should be noted that the higher number of methods and techniques used within AMT projects evaluation process should be facilitated by various pieces of software. Therefore, companies were asked if spreadsheet packages, dedicated software or other computer aids were used in the process of evaluating advanced manufacturing technology investment proposals and the results are shown in table 7. We can see that a very high proportion of Czech companies use spreadsheet software (75.7 % in 1999 and even 89.4 % in 2005). Approximately one out of six managers employs some dedicated computer software, while other computer aids were reported to be used semi-occasionally.

Table 7: Use of computer aids

Computer aids	1999 [%]	2005 [%]
Spreadsheets	75.7	89.4
Dedicated software	16.2	18.3
Other computer aids	4.1	4.8

Of course, it was anticipated that conventional criteria are widely used and therefore the respondents were asked to indicate, based on their own experience and judgment, whether or not they agreed with the statement that, "conventional appraisal methods such as Payback, NPV and IRR favored short term projects". According to (Lefley & Sarkis, 1997) more than 70 % of companies in the UK and USA agreed with the statement, while significantly fewer in the Czech Republic (55.6 % in 1999 and 53.2 % in 2005) were of the same opinion. The relatively low proportion of Czech managers who thought conventional techniques favor short term investments seems to support the above mentioned views that conventional financial appraisal methods do not automatically favor short-term projects and that these criteria could be used for AMT project proposal evaluation too. Of course, these techniques should be used wisely because when short payback periods or unjustifiably high discount rates are used then a short-term bias can easily occur. In this respect it could be interesting to find out if there is a tendency to set up some tight hurdle rates for AMT projects justification in companies. The respondents were asked to express their level of agreement with the relevant statement and their responses are summarized in table 8. We can see that nearly every second manager agreed with the statement and admitted that there is a tendency to set up very tight hurdle rates which could indicate rather disadvantageous starting position for AMT projects. High hurdle rates in combination with the above mentioned traditional appraisal methods could easily result in the AMT project rejection. On the other hand we have to say that in many cases the high hurdle rates are used by managers in order to make appropriate adjustment for a higher degree of risk and uncertainty that relates to AMT projects and it is rather typical approach taken by many companies worldwide when evaluating more risky investment project. Hence such behavior should not be automatically perceived as deliberate intention to discriminate against AMT projects especially when the particular company lacks experience with the project proposal that is under consideration.

Table 8: There is a tendency to set too high hurdle rates for AMT projects

There is a tendency to set too high a hurdle rate for AMT projects.	1999 [%]	2005 [%]
Agree	48.0	48.8
Disagree	52.0	51.2

Some researchers as well as practitioners advocate for exploitation of non-financial criteria and rather strategically oriented criteria believing that there is too much importance attached to conventional techniques. That is why the respondents were asked to express, based on their own experience and judgment, whether or not they agreed with the statement that, "too much importance is attached to conventional techniques". Their responses are presented in table 9.

Table 9: Too much importance is attached to conventional techniques

Too much importance is attached to conventional techniques	1999 [%]	2005 [%]
Agree	51.4	44.4
Disagree	48.6	55.6

It is clear that Czech managers do not feel like having a serious problem with conventional appraisal techniques utilization and their views are perfectly conformable with the opinions of British and US managers where also slightly less than five out of ten managers agreed with the above presented statement that too much importance is attached to conventional appraisal techniques. It was also noted (Hynek & Janeček, 2007) that a high proportion of companies in all three countries (83.4 % on an average) referred back for re-appraisal those proposals that had failed the initial financial appraisal (the results concerning the situation in the Czech Republic are displayed at table 10). Of course, the introduction of a referral process into the investment justification procedure creates further opportunity for managers to examine the whole proposal carefully once more, to take into account strategic considerations, re-assess and quantify potential benefits or even adjust financial criteria that has to be fulfilled (for example, by reduction of required payback period, or by lowering the pertinent discount rates). On the other hand, it is the very same moment when exactly opposite measures and actions could be taken and there is a large space in which the accept/reject decision could be manipulated. It could be anticipated that in these cases the formal appraisal procedure transforms itself into a ritual where the final decision is based on other influences, which might be of a political, rather than an economic nature. In this context we should put and understand the interesting fact that more than eight out of ten respondents confirmed the referral procedure.

Table 10: Percentage of proposals re-appraised

Project proposals re-evaluated	1999 [%]	2005 [%]
Agree	89.2	81.5
Disagree	10.8	18.5

Having admitted that there might be some political influence in the referral procedure and in the projects evaluation process in general it is natural to ask to which extent do senior executives use their dominant role based on their formal as well as informal authority in order to affect the relevant decisions related to AMT investment in both directions. That is the reason why the respondents were asked to express their level of agreement with the statement that more importance is attached to the experienced judgment of senior management than to financial indicators. The results are shown in table 11 and we can see that slightly over fifty percent of Czech respondents agreed with the statement in 1999 (51.9 %) and their number declined further in 2005 (45.7 %). Nevertheless, it should be stressed that the number of managers who agreed with the statement is relatively high overall and it is clear the concerns expressed by some researchers as well as practitioners seems to be legitimate.

Table 11: More importance is attached to the experienced judgment of senior management than to financial indicators

More importance is attached to the experienced judgment of senior management than to financial indicators	1999 [%]	2005 [%]
Agree	51.9	45.7
Disagree	48.1	54.3

To conclude this section we would like to stress that despite the mentioned criticism the traditional financial appraisal techniques play important role in the process of AMT projects evaluation and justification. We have shown that managers prefer the simplest techniques like payback period that are very easy to understood and interpret, but we have also mentioned that there is a danger of bias towards projects delivering short-term profits when these techniques are used mechanically, shortsightedly, and without broader impact considerations. Furthermore, the risk that AMT projects would be disadvantaged by utilization of these simple techniques could be moderated by utilization of several methods and we have shown that more than eight out of ten projects are re-evaluated if they failed to pass through the initial financial evaluation process.

Personal Opinions of Managers

We have already mentioned that the process of AMT justification might be seriously influenced by management attitudes towards technology investment in general. AMT is often considered as one of critical factors that plays important role in the process of acquiring competitive advantage. It seems to be a widely accepted opinion but do managers really think so? We wanted to verify this ascertainment and the respondents were asked to indicate based on their own experience and judgment, whether or not they agreed with the statement that non-investment in AMT was a high risk strategy. Responses to this statement were summarized in table number 12 and we can immediately see there that surprisingly large proportion of executives in the Czech Republic

(33.3% in 1999 and 30.7 % in 2005) disagreed that non-investment in AMT is a high-risk strategy.

Table 12: Non-investment in AMT is a high-risk strategy

Non-investment in AMT is a high-risk strategy	1999 [%]	2005 [%]
Agree	66.7	69.3
Disagree	33.3	30.7

Comparing these findings with the results of Lefley and Sarkis (1997) who reported more than the decade ago than 74.8 % in the UK and 81.9 % in the US agreed with the statement that non-investment in AMT is a high-risk strategy, it is clear that significantly higher proportion of Czech managers do not consider AMT as strategically important investment. It is a rather surprising ascertainment taking into account that Czech manufacturing companies after transformation of our economy had to find new market opportunities for their products. Many of them oriented themselves mainly on strongly competitive markets in Western Europe, many others were sold to foreign investors and it was anticipated that new owners would bring new technologies too. It is difficult to generalize, but as we concluded in (Hynek & Janeček, 2006a) it is likely that many companies have apparently decided to rely on skilful and relatively cheap labor force and that is why the relevant companies seems to be somewhat slow in AMT adoption. Moreover, we are afraid that in today's mutually interlinked and quickly changing global world the exploitation of such strategy sounds like a rather shortsighted decision. Obviously it is not easy to change management attitudes towards AMT investment and perception of its importance from day to day. Fortunately enough, there are some other issues we should pay our attention too and we think that there might be some space where improvement of the current state of art is more feasible. Moreover, we will show that while there are significant differences in perception of the strategic importance of AMT investment in general between managers working under conditions of transforming Central European economy and managers representing two of the most developed countries in the world, there are some problems they have in common too. For example, we have learned that many AMT projects are likely to be rejected just because the lack of understanding of what the contribution of new technology really is. We could see in table number 13 that more than 60 % of Czech executives agree with the statement that it is difficult to assess all potential benefits of AMT investments (67.1 % in 1999 and 60.3 % in 2005). The level of agreement with the relevant statement was even higher in the UK (81.6 %) while 63.9 % of American managers shared the view (Lefley & Sarkis, 1997). Thinking about reasons that we can see three possible explanations of this unfavorable situation First of all, there are some benefits where managers seems to be unable to foresee and to assess their impact and magnitude there because of lack of experience, lack of relevant input data etc. Secondly, the company is not sure whether some particular benefit will be realized at all and thus the benefit falls into this category and stays there without any attempt to quantify it. And thirdly, it is often believed that brand new technology will bring along some new benefits and completely unexpected benefits that are impossible to predict before the technology reach the stage of regular utilization. While the first problem seems to be based on lack of experience and administrative-technical reasons, the other two explanations seems to be much more of a speculative nature.

Table 13: AMT investments are difficult to assess

AMT investments are difficult to assess because they have non-quantifiable benefits	1999 [%]	2005 [%]
Agree	67.1	60.3
Disagree	32.9	39.7

However, whatever reason applies it helps to create the feeling that there are some non-quantifiable benefits that were not taken into account. And we will demonstrate that there is a problem related to proper assessment of non-quantifiable benefits and their expression in financial terms which means that these benefits will not be taken into relevant economic calculations. We can see in table 14 that large majority of Czech managers (90.1 % in 1999 and 81.7 % in 2005) agreed with the statement that not all potential benefits of AMT are taken into account because they are difficult to quantify in financial terms. It should be noted that these numbers are in compliance with the earlier findings of (Lefley & Sarkis, 1997) who reported that 80.9 % of British managers agreed with the statetems and 81.2 % of American managers did so. It is important to repeat here that the respondents of our surveys were financial directors and decision makers of surveyed manufacturing companies. Recalling back this fact we can see that the situation is very serious and some measures should be taken in order to make sure that AMT project proposals have a fair chance to get through the justification process and to get the pertinent investment approval.

Table 14: Not all benefits are taken into account

Not all potential benefits of AMT are taken into account because they are difficult to quantify in financial terms	1999 [%]	2005 [%]
Agree	90.1	81.7
Disagree	9.9	18.3

According to Primrose (1991) people advocating investment in AMT have made considerable efforts to identify the company-wide benefits which it can produce. The problem is that they describe these benefits always in general terms, such as the following: increased flexibility of production, better-quality products, improved documentation, ability to respond to market needs, need to keep up with competition, improved company image, better management control, obtaining experience of new technology, etc. Managers usually start with the belief that a particular aspect of AMT could be used in their department and they would select an application which was aimed at improving operating efficiency. Having defined the required specification, they try to justify the expenditure afterwards. And now it is necessary to identify the benefits. The nature of intangible benefits is such that they do not have to appear in the department where the investment is made, but occur elsewhere in the company. In addition, the relationship between cause and effect is indirect, so that their magnitude has to be estimated rather than directly calculated. In fact there are two distinct problems and these must be dealt with separately. First of all the form in which the benefit is quantified, and secondly estimating the magnitude of the benefit (see (Primrose, 1991) for more details).

CONCLUSIONS

The presented selected results of two AMT surveys focused on the specific issues of advanced manufacturing technology justification that were carried out in the Czech Republic demonstrate that the economic justification of the relevant projects is definitely not an easy process. Moreover, there are many problems that seem to be common for managers in Central Europe who has to face the conditions of transforming economy and managers from technologically most developed economies in the world. First of all, our results clearly demonstrate that Czech manufacturing companies are lagging behind their British and American competitors in the adoption of AMT and the optimistic prognoses that the high level of foreign direct investment will bring along acceleration of AMT adoption as well as the latest technology has not been proved yet. We have also shown some pieces of evidence that AMT projects might be very easily knowingly as well as unknowingly disadvantaged because of a whole spectrum of reasons. Based on our results it is clear that managers exploit rather unsuitable financial criteria, too much importance is given to the simplest methods that clearly prioritize short-term outcomes and thus short-term projects. British and American managers seem to be more aware of this fact and perhaps it is the reason why they tend to utilize more sophisticated criteria and greater number of criteria in general than managers in the Czech Republic do. However, we have stressed that the problems could be avoided if the criteria are used wisely and we have mentioned as an example the difference between payback period utilization in western companies on one side, and Japanese companies on the other one. We have seen that more than eight of out ten AMT projects are re-evaluated if they failed the initial financial appraisal. As the result of this phenomenon many projects are carried out only partially. It could be the restricted version of the original project that lacks the originally intended level of integration, or it could be done at the expense of the originally planned level of technology used. In both cases there is a danger that restricted version of the originally planned AMT project will be unable to deliver originally planned benefits and the project will not live up the expectations. Furthermore, we have pointed out that introduction of the referral process establishes ground for various influences that might be of a political rather than economic nature. Finally, we have examined management attitudes towards AMT projects. We have realized that comparing our results with the outcomes of earlier survey conducted in the UK and USA, significantly higher proportion of Czech managers do not consider AMT as strategically important investment. On the other hand, there are some serious issues that significantly influence the process of AMT adoption and these issues are common for the managers from all three surveyed countries. First of all, two thirds of managers agreed with the statement that AMT investments are difficult to assess because they have nonquantifiable benefits. Secondly, over eighty percent of respondents supported the view that not all potential benefits of AMT are taken into account because they are difficult to quantify in financial terms. Putting these ascertainments in other words we can see that there is a clear lack of understanding of what the contribution of the proposed AMT project really is. Moreover, managers are fully aware of the fact that some benefits are not taken into their calculations because they are unable to estimate them and express them in financial terms. We have already expressed (Hynek & Janeček, 2006b) our view that there is an important space and great opportunity right here that should be taken by technology specialists. They should be able to identify, describe and explain the complex benefits of a particular AMT project and hereby prepare better background material for financial executives. Their involvement in this phase could assure that various tangible as well as intangible benefits will be taken into consideration, properly assessed and consequently expressed in financial terms. Of course, this task could be fulfilled only by technology experts who are able to see the particular technology not simply from technological point of view. Their knowledge and broader understanding of technology benefits for the company as a whole could considerably improve the chances of AMT projects to get the management approval. Of course, it should be also accentuated that economic approach to AMT projects justification is widely used but it is not the single approach and we recommend employing strategic and analytical approaches too. These approaches do have their own drawbacks too and therefore wise combination of different approaches should be encouraged in order to make sure that AMT project proposals are assessed properly. This is the only way providing enough opportunities to avoid later disappointment.

ACKNOWLEDGEMENT

This research has been supported by the Grant Agency of the Czech Republic project No. 402/07/1495.

REFERENCES

1. Abdel-Kader, M. G. & Dugdale, D. (1998). Investment in Advanced manufacturing technology: a study of practice in large U.K. companies. Management Accounting Research, No. 9, pp. 261-284, ISSN 1044-5005.

2. Ariss, S. S.; Raghunathan, T. S. & Kunnathar, A. (2000). Factors Affecting the Adoption of Advanced Manufacturing Technology in Small Firms. SAM Advanced Management Journal, Vol. 65, No. 2, Spring 2000, pp. 14-29, ISSN 0749-7075.

3. Bucher, P.G. & Lee, G.L. (2000). Competitiveness Strategies and AMT Investment Decisions. Integrated Manufacturing Systems, No. 11/5, pp. 340-347, ISSN 0957-6061.

4. Chan, F.T.S.; Chan, M.H.; Lau, H. & Ip, R.W.L. (2001). Investment Appraisal Techniques for Advanced Manufacturing Technology (AMT): A Literature Review. Integrated Manufacturing Systems, No. 12/1, pp. 35-47, ISSN 0957-6061.

5. DeRuntz, B. D. & Turner, R. M. (2003). Organizational Considerations for Advanced Manufacturing Technology. International Journal of Production Economics, Vol. 2002, No. 79, pp. 197-208, ISSN 0925-5273.

6.   Dornan, S. B. (1987). Cells and Systems: Justifying the Investment. Production, February 1987, pp. 30-35.

7.   Hofmann, C. & Orr, S. (2005). Advanced Manufacturing Technology Adoption – the German Experience. Technovation, Vol. 25, No. 7, pp. 711-724, ISSN 0166-4972.

8.   Hynek, J. & Janeček, V. (2005). Adoption of Advanced Manufacturing Technology – New Trends in the Czech Republic. In: Proceedings of the IEEE 9th International Conference on Intelligent Engineering Systems. IEEE, Piscataway, NJ, 2005, pp. 75-78, ISBN 0-7803-9474-7.

9.   Hynek, J. & Janeček, V. (2006a). Information Gap between Technology Specialists and Decision Makers. In: Proceedings of IEEE 3rd International Conference on Mechatronics, IEEE, Piscataway, NJ, pp. 61-64, ISBN 1-4244-9712-6.

10. Hynek, J. & Janeček, V (2006b). Problems of Advanced Manufacturing Technology Projects Approval. In: Proceedings of the 5th WSEAS Int. Conf. on System Science and Simulation in Engineering (ICOSSE'06), Tenerife, Canary Islands, Spain, WSEAS Press, pp. 412-416, ISBN 960-8457-57-2, ISSN 1790-5117.

11. Hynek, J. & Janeček, V. (2007). Advanced Manufacturing Technology Projects Justification. In: Proceedings of the 4th IEEE International Conference on Mechatronics, University of Kumamoto, Japan, 2007, pp. 277-282, ISBN 1-4244-1184-X.

12. Hynek, J. & Janeček, V. (2008). Economic Justification of Advanced Manufacturing Technology, In: Proceedings of the 2nd WSEAS Int. Conf. on Management, Marketing and Finances (MMF'08), Harvard University, Cambridge, Massachusetts, WSEAS Press, 2008, pp. 103-108, ISSN 1790-5117.

13. Lefley, F. & Sarkis, V. (1997). Short-termism and the appraisal of AMT capital projects in the US and UK. International Journal of Production Research, Vol. 35, No. 2, pp. 341-368,ISSN 0020-7543.

14. Lefley, F. & Wharton, F. (1993). Advanced Manufacturing Technology Appraisal: A Survey of U.K. Manufacturing Companies. Proceedings of the 4th Int. Production Management Conference: Management and New Production Systems, London Business School, 1993, pp. 369–381.

15. Lefley, F. (1994). Capital investment appraisal of advanced manufacturing technology. International Journal of Production Research, Vol. 32, No. 12, pp. 2751-2776, ISSN 0020- 7543.

16. Lefley, F.; Wharton, F.; Hájek, L.; Hynek, J. & Janeček, V. (2004). Manufacturing investments in the Czech Republic: An international

comparison. International journal of Production Economics. Vol. 88, No. 1, pp. 1-14, ISSN 0925-5273.

17. Meredith, J. R. & Suresh, N. (1986). Justification techniques for advanced manufacturing technologies. International Journal of Production Research, Vol. 4, No. 5, pp. 1043-1058, ISSN 0020-7543.

18. Primrose, P. L. (1991). Investment in Manufacturing Technology. Chapman&Hall, London 1991, ISBN 0412409208.

19. Raafat, F. (2002). A comprehensive bibliography on justification of advanced manufacturing systems. International Journal of Production Economics, Vol. 2002, No. 79, pp. 197-208, ISSN 0925-5273.

20. Small, M. H. (2006). Justifying Investment in Advanced Manufacturing Technology: a Portfolio Analysis. Industrial Management & Data Systems, Vol. 106, No. 4, pp. 485- 508, ISSN 0263-5577.

Chapter 12

FLEXIBLE MANUFACTURING SYSTEM SIMULATION USING PETRI NETS

Carlos Mireles, Alfonso Noriega and Gerardo Leyva

University Campus STeP Ri Slavka Krautzeka 83/A 51000 Rijeka, Croatia

INTRODUCTION TO PETRI NETS

In 1962 German mathematician Kart Adam on his Ph.D. work "Kommunikation mit automaten", he proposed a new way for modelling & representating systems where events and transitions are present. This new representation is now known as Petri Nets (PN). Since then, several researchers have used this tool. In the USA and in Europe several developments have been done. In 1980, in Europe, a work table was controlled using PN and the related work was presented at one International Conference. Researchers in France have done great contributions to the development and the applications of PN. Some of them used the PN to describe Programmable Logic Controllers (PLC) and this application had a big influence in the development of the Grafcet.

Petri Nets Theory

Basic Concepts

Petri Net

Graphic and executable technique to specify and analize dynamic & concurrent discrete event systems

Formal

Petri Nets analysis is a mathematical technique well defined. Many static and dynamic properties of a PN (and therefore for a system represented by a PN) can be mathematically proved.

Graphics

This technique belongs to the area of mathematics known as "Graph Theory", which makes a PN being able to represented both by a graphic and by mathematic expressions. This graphic property provides a better understanding of the system which is represented by the PN. Petri Nets analysis is a mathematical technique well defined. Many static and dynamic properties of a PN (and therefore for a system represented by a PN) can be mathematically proved.

Graphics

This technique belongs to the area of mathematics known as "Graph Theory", which makes a PN being able to represented both by a graphic and by mathematic expressions. This graphic property provides a better understanding of the system which is represented by the PN.

Executable

A PN can be executed and then the dynamics of the system can be observed.

Concurrency

Multiple independent entities can be represented and supported by PN.

Discrete event Dynamic Systems

A system that can change its current state based in both its current state and the transition conditions between states.

Structure of a Petri Net

Formal Definition

A Petri Net has a set of Places, a set of Transitions, an Input Function and an Output Function. The structure of a Petri Net is the array (L, T, E, S, m0) where:

- L is the set of places in the graph.
- T is the set of transitions in the graph. L & T satisfy the following conditions $L \cup T = \emptyset$ $L \cap T = \emptyset$.
- $E: L x T \rightarrow \{0,1\}$ is the input function which specify the connectivity between Places & Transitions.
- $S: L x T \rightarrow \{0,1\}$ is the output function which specify the connectivity between Transitions & Places.

Graphic Representation

A Petri Net is an oriented graph that contains two types of nodes: Places and Transitions which are connected by oriented arcs that connect Places with Transitions –connectivity between nodes of the same kind is not allowed. In the graphic representation, Places are shown as circles, Transitions are shown as bars and the arcs are shown as arrows. Places represent conditions and Transitions represent events. A Transition has certain number of places either as inputs or outputs which are pre and post conditions. See figure 1.

Marking a Petri Net

The marking of a Petri Net is a positive integer μ_i for every place Li. A mark is represented as a dot within the circle for a given place. These marks move between the places which provide the dynamic feature of the Petri Net. A Petri Net is considered marked when at least one place has a mark. One place can have N marks, where N is a positive integer. If N = 0 then the place has no marks. A marker M is a function M L: $\rightarrow$N that can be expressed by:

$$M = \begin{bmatrix} m_1 \\ m_2 \\ \vdots \\ m_N \end{bmatrix}$$

(1)

m_i place L_i is the number of dotsfor .

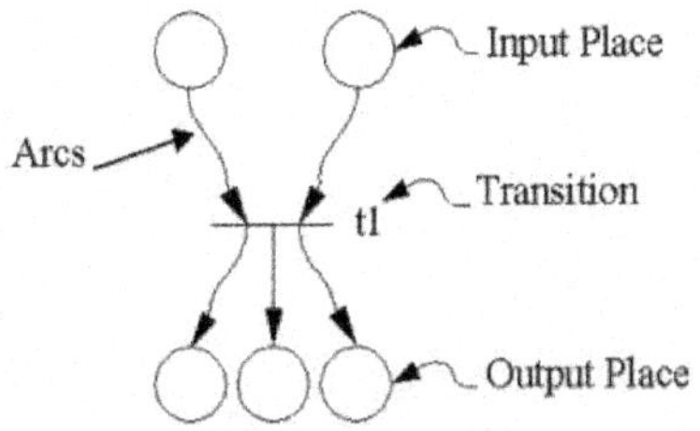

Figure 1: Graphic representation of a PN showing the Input & Outputs places as well as the Transition.

Interpretation of a Petri Net

Marks are the resources. Resources can be physical entites or no physical such as messages or information. Places are the locations where the resources are stored. Transitions are actions that transfer the resources to generate new resources.

The weight of each arrow (a number on each arrow) is the minimum number of resources needed in a place in order to get the action indicated by the transition.

Triggering a Transition in a Petri Net

To trigger a Transition, it has to be validated which means that every Place connected to this Transition has to have at least one mark. The execution of a Petri Net is controlled by the number and distribution of the marks on their Places. When a Transition is triggered, a change in the marking of the Petri Net is performed and every input place lose the number of marks indicate by the weight in the arc connecting the Transition with that input place. On the other hand, every output place gain the number of places inidicated by the weight in the arc connecting the Transition with that output place. In figure 2 are shown different cases for the triggering of a transition.

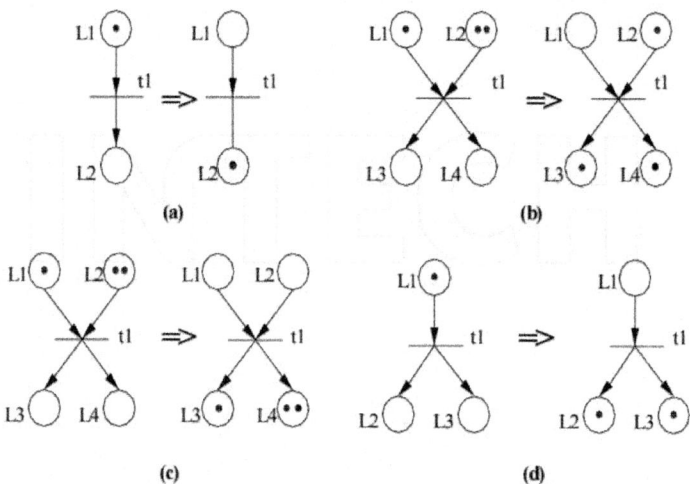

Figure 2: Different triggering cases for some Petri Nets.

PETRI NETS SHOWING RELATIONSHIPS AMOG SUB-PROCESSES.

Task Sequencing

Figure 3 shows how different task can be done in sequential way.

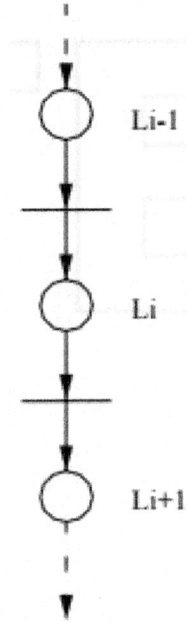

Figure 3: Task Sequencing

Task Selector

Figure 4 shows a task selector. This graph is useful when we need in a system an EXOR function between transitions t_1 & t_2.

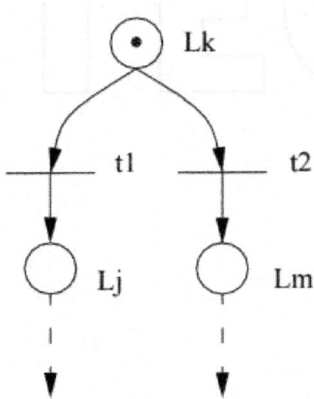

Figure 4: Task Selector

Task Synchronization

This is a very useful concept to synchronize two or more ub-processes assuming that they may have different running times or different evolution in time. Any of the processes can be out of phase respect to the others, however this model provides a wait state for any of the processes until the triggering of the transition. Synchronization is achieved when, being validated by the proper marking on the input places, the event occurs triggering the transition.

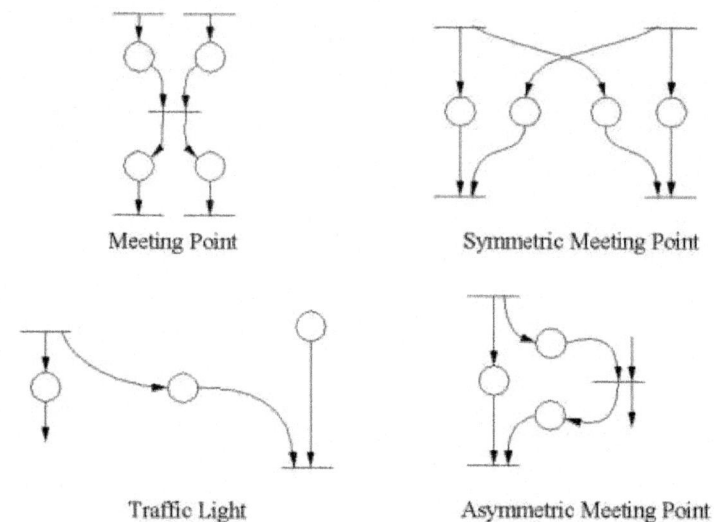

Meeting Point   Symmetric Meeting Point

Traffic Light   Asymmetric Meeting Point

Figure 5: Different models to synchronize processes

Tasks Concurrency

Figure 6 shows how concurrent processes are represented.

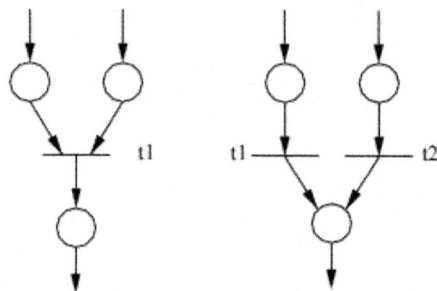

Figure 6: Tasks Concurrency

Sharing Resources

Sharing resources is done when two or more sub-processes are using a unique resource in the system –such a robot to load/unload two conveyors. The objective is to avoid the malfunction of one of the processes because the resource is not available.

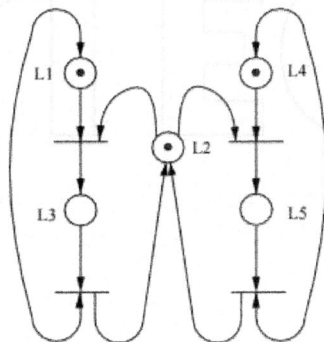

Figure 7: Sharing Resources

EQUATION OF A PETRI NET

Figure 8 shows a Petri Net which will be used to explain the Petri Net equation.

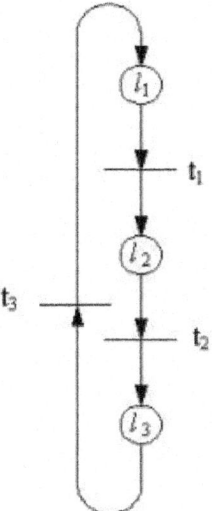

Figure 8: Sharing Resources

Input Matrix

This matrix provides the connectivity information from Places to Transitions. In this case Place l_1 goes to Transtion t_2, Place l_2 goes to Transition t_2 and so on. See figure 8.

$$E = \begin{array}{c} \quad l_1 \; l_2 \; l_3 \\ \begin{bmatrix} 1 & 0 & 0 \\ 0 & 1 & 0 \\ 0 & 0 & 1 \end{bmatrix} \begin{array}{c} t_1 \\ t_2 \\ t_3 \end{array} \end{array}$$

(2)

Output Matrix

This matrix provides the connectivity information from Transitions to Places. In this case Transtion t2 goes to Place l_1, Transtion t_2 goes to Place l_2 and so on. See figure 8.

$$S = \begin{array}{c} \quad t_1 \; t_2 \; t_3 \\ \begin{bmatrix} 0 & 0 & 1 \\ 1 & 0 & 0 \\ 0 & 1 & 0 \end{bmatrix} \begin{array}{c} l_1 \\ l_2 \\ l_3 \end{array} \end{array}$$

(3)

General Matrix of a Petri Net

The two last matrixes are used to define the Net Matrix given by:

$$A = S - E$$

(4)

Evolution of a Petri Net

The evolution of the Petri Net is related with the "movement" of the marks in the graph from one place to another. The recursive equation is given by:

$$\mu_{k+1} = \mu_k + AV_k$$

(5)

- μ_k represents the marking vector before the evolution of the net.
- μ_{k+1} represents the marking vector after the evolution of the net.
- A represents the General Matrix of the net.
- V_k represents the Transitions that are been triggered in the current evolution step   of the net.

A working Petri Net

Figure 9 shows a Petri Net before and after the evolution step (trigger of t_1).

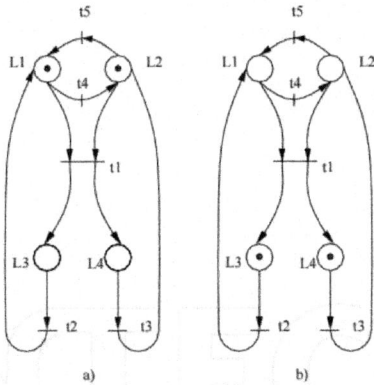

Figure 9: A Petri Net before and after the trigger of transition t$_1$

APPLICATION EXAMPLES

Wagon Control

Figure 10 shows a wagon that moves along a rail by the action of signals R (right) and L (left). Extreme positions are detected by sensors A & B. Button M starts the sequence when wagon is in initial position.

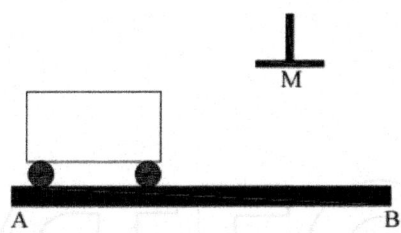

Figure 10: Wagon Control

There are three places (l$_1$, l$_2$ & l$_3$): Idle, Moving from A to B & Moving from B to A. There are three transitions (t$_1$, t$_2$ & t$_3$): Switch M, wagon detected by sensor A & wagon detetcted by sensor B. When M is press, the wagon moves from A to B, once reaches this point goes back to A. Sequence can only starts when wagon is in Idle again.

Figure 11 shows the Petri Net for this system.

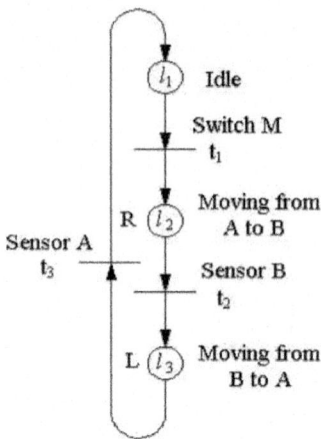

Figure 11: Petri Net for the Wagon Control system

Following the matrixes for this system:

$$E = \begin{bmatrix} \overset{t_1}{1} & \overset{t_2}{0} & \overset{t_3}{0} \\ 0 & 1 & 0 \\ 0 & 0 & 1 \end{bmatrix} \begin{matrix} l_1 \\ l_2 \\ l_3 \end{matrix} \quad S = \begin{bmatrix} \overset{t_1}{0} & \overset{t_2}{0} & \overset{t_3}{1} \\ 1 & 0 & 0 \\ 0 & 1 & 0 \end{bmatrix} \begin{matrix} l_1 \\ l_2 \\ l_3 \end{matrix} \quad V_0 = \begin{bmatrix} 1 \\ 0 \\ 0 \end{bmatrix} \quad \mu_0 = \begin{bmatrix} 1 \\ 0 \\ 0 \end{bmatrix}$$

(6)

$$A = S - E$$
$$A = \begin{bmatrix} -1 & 0 & 1 \\ 1 & -1 & 0 \\ 0 & 1 & -1 \end{bmatrix}$$

(7)

Where $\mu_{k+1} = \mu_k + AV_k$ then:

$$\mu_1 = \mu_0 + AV_0 : \mu_1 = \begin{bmatrix} 1 \\ 0 \\ 0 \end{bmatrix} + \begin{bmatrix} -1 & 0 & 1 \\ 1 & -1 & 0 \\ 0 & 1 & -1 \end{bmatrix} \cdot \begin{bmatrix} 1 \\ 0 \\ 0 \end{bmatrix} = \begin{bmatrix} 1 \\ 0 \\ 0 \end{bmatrix} + \begin{bmatrix} -1 \\ 0 \\ 0 \end{bmatrix} = \begin{bmatrix} 0 \\ 1 \\ 0 \end{bmatrix}$$

(8)

$$\mu_1 = \mu_0 + AV_0 : \mu_1 = \begin{bmatrix} 0 \\ 1 \\ 0 \end{bmatrix} + \begin{bmatrix} -1 & 0 & 1 \\ 1 & -1 & 0 \\ 0 & 1 & -1 \end{bmatrix} \cdot \begin{bmatrix} 0 \\ 1 \\ 0 \end{bmatrix} = \begin{bmatrix} 0 \\ 1 \\ 0 \end{bmatrix} + \begin{bmatrix} 0 \\ -1 \\ 0 \end{bmatrix} = \begin{bmatrix} 0 \\ 0 \\ 1 \end{bmatrix}$$

(9)

$$\mu_1 = \mu_0 + AV_0 : \mu_1 = \begin{bmatrix} 0 \\ 0 \\ 1 \end{bmatrix} + \begin{bmatrix} -1 & 0 & 1 \\ 1 & -1 & 0 \\ 0 & 1 & -1 \end{bmatrix} \cdot \begin{bmatrix} 0 \\ 0 \\ 1 \end{bmatrix} = \begin{bmatrix} 0 \\ 0 \\ 1 \end{bmatrix} + \begin{bmatrix} 1 \\ 0 \\ -1 \end{bmatrix} = \begin{bmatrix} 1 \\ 0 \\ 0 \end{bmatrix}$$

(10)

SIMULATION SOFTWARE

In order to simulate the Petri Nets, a software kit was developed. Although there are some simulators available, we took the decision to develop our own as the main goal and to have a software system able to be interfaced with the real world. The software was developed in Lab Windows CVI, having a Graphic User Interface which allows the user to create easily the net that represents the system that is needed to simulate. Figure 12 shows the tipycal interface of the system.

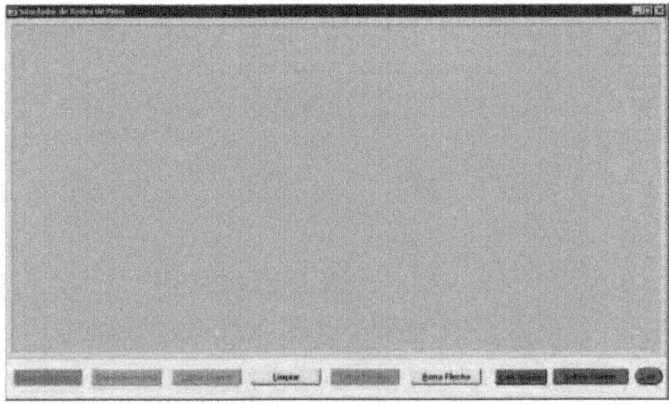

Figure 12: Main view of the simulator

The operator inputs the data related with the PN structure, the information is then interpreted by the simulator which shows the development of the PN. On figure 13, the Petri Net shown on figure 11 was introduced to the simulator.

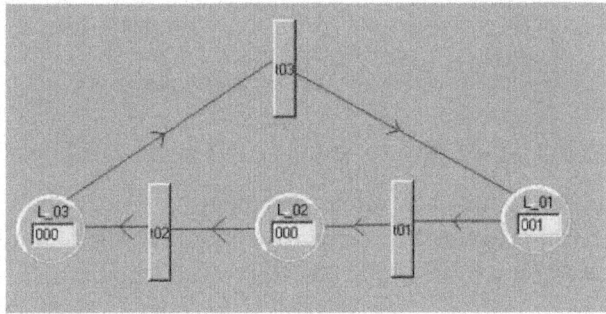

Figure 13: Petri Net for the Wagon Control system

Figures 14, 15 & 16 shows the status of the Petri Net when transitions t_1, t_2 & t_3 are triggered

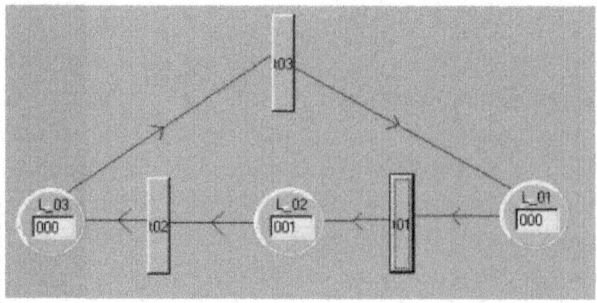

Figure 14: Petri Net for the Wagon Control system after t_1 is triggered

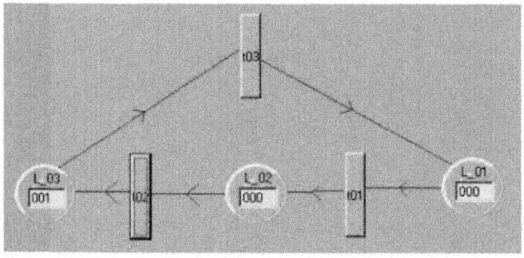

Figure 15: Petri Net for the Wagon Control system after t_2 is triggered

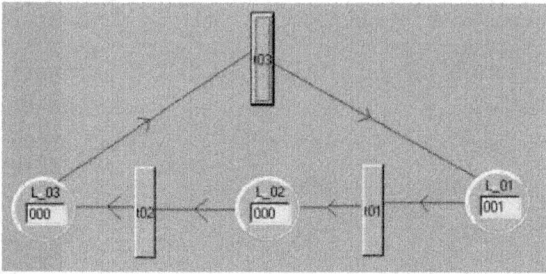

Figure 16: Petri Net for the Wagon Control system after t_3 is triggered

Inside the simulator each circle represents a Place where the wagon is moving to the right, to the left, or simply is on the idle state waiting for the button M to be pushed (transition t_1) to start a sequence.

It is worth to say that this simulator can easily be used as a Control System just by adding an Acquisition Board for sampling the actual signals which in turn can activate the transitions. The purpose with this was to develop a non expensive control system which can be used by very small companies that do not have the capital to buy very expensive equipment. In comparison with a PLC controller the cost involved when implementing this approach is very low and also offers an intuitive way of developing the control system.

Benefits of a Petri Net

Petri Nets have the following benefits:

- They have a graphic representation which allows a simply and secure way to observe how the system is working.
- They have semantics well defined that specify the net.
- They can be used to represent several types of systems.
- They have an explicit description for states and actions.
- They provide interactive simulations where the results are shown directly on the graphic diagram of the net.

In future research this system can be used on the control of production lines where series of events (discrete events) need to be controlled, such as a robot feeding a conveyor, a conveyor moving cans or packages into storage areas, etc.

REFERENCES

1. Mireles, C. et al. (2004), Flexible Manufacturing System Simulation using Petri Nets

Chapter 13

ENHANCING MULTISTAGE DEEP-DRAWING AND IRONING MANUFACTURING PROCESSES OF AXISYMMETRIC COMPONENTS: ANALYSIS AND EXPERIMENTATION

F. Javier Ramírez,[1,2] Rosario Domingo,[3] Michael S. Packianather,[4] and Miguel A. Sebastian[3]

[1]EXPAL Systems, Avenida Partenón 16, 28042 Madrid, Spain

[2]School of Industrial Engineering, University of Castilla-La Mancha, Albacete, Spain

[3]School of Industrial Engineering, UNED, 28042 Madrid, Spain

[4]School of Engineering, Cardiff University, Cardiff, UK

ABSTRACT

An optimization technique for combined processes of deep-drawing and ironing has been created in order to improve the total process time and cost in manufacturing procedures of axisymmetric components. The initial solution is optimized by means of an algorithm that minimizes the total time of the global process, based on relationship between lengths, diameters, and velocities of each stage of a multistage process and subject to constraints related to the drawing ratio. The enhanced solution offers a significant reduction in time and cost of the global process. The final results, applied to three cases, are compared with experimental results, showing the accuracy of the complete solution.

INTRODUCTION

The industry of metallic components manufacturing requires developments to be more efficient, in particular in the deep-drawing procedures, where it is important to decrease the process times and costs as in mass production. Thus, it is necessary to devise specific algorithms that will satisfy these demands. These algorithms should be based on technological and scientific basis that will provide solutions that are ready for transfering to the industries. The deep-drawing process has been analysed with this objective in mind due to its convenient nature as a global model which includes all stages of the process, namely, drawing, redrawing, and ironing.

The majority of literature contributions are focused on the study of properties of process, in particular, the prediction of the limiting drawing ratio (LDR) [1–3], the blank design using different methodologies, such as parametric NURBS surfaces [4], upper bound method [5], or artificial neural network [6, 7], the effect of die radius on the blank holder force and drawing ratio [8], the predicted thickness distribution of the deep drawn circular cup of stainless steel [9], the improvement of drawability by means of technological parameters [10, 11] or the formability with different thickness [12]. However, some efforts have been realised about the parts design [13] or generation of algorithms, mainly related to the process planning; Ramana and Rao [14] developed a framework based on knowledge related to design-process planning integration for sheet metal components, although there is no evidence of its application. Also, Vosniakos et al. [15] devised an intelligent system to process design of sheet parts. As can be seen, the researches of deepdrawing processes are not focused on the reduction of time, despite frequently being used on mass production due to the characteristics of the parts

This paper presents a model that provides a comprehensive analysis of those phenomena occurring in the multistage processes of axisymmetric geometry and applied to the manufacturing of this type of components. The scientific development stems from the work done by Leu [1] and Sonis et al. [2] which provides LDR solutions based on normal anisotropy value, strain hardening exponent, and others, applied to the drawing and redrawing stages. The model [16, 17] has a scientific foundation based on the literature covering the plastic deformation processes and, in particular, drawing [18], redrawing [19], and ironing. This study focuses on analyzing geometries of axisymmetric components, manufactured by a multistage deep-drawing process.

The drawing processes from a blank [6, 8] have been researched and their governing equations are well known [3, 5]. However, the redrawing and ironing phenomena have received relatively less attention [2, 20]. This work contributes to the definition of a global process of drawing, redrawing, and ironing combined process, together with a common focus on the drawing process in order to obtain a global solution.

The model is used to carry out a quantitative and integrated analysis in a multistage deep-drawing process based on a scientific solution and real measurements. Also, the model permits the modification for some process variables to predict their influence on the process [9]. In addition, it is based on the definition of limiting conditions to guarantee the stability of the process. The simultaneous accomplishment of limiting situations of each process (drawing, redrawing, and ironing) allows for fixing a boundary for the values of each stage which gives the initial solution. The model permits the optimisation of the initial solution from several points of view: total process time, manufacturing cost, among others, such as punch and die wear. This optimisation is based on the resolution of an algorithm by means of recursive functions, which explores all the possibilities of the process and selects the most adequate one. The algorithm is supported by a software tool that provides the optimized solution [21]. The optimized solution is compared with the experimental results.

The remainder of the paper is organised as follows. In Section 2, the methodology used is formulated and applied to some industrial cases. In Section 3, the initial solution found from the scientific model is presented. Section 4 defines the optimization process. In Section 5, the optimization algorithm for solving the multistage process is proposed. In Sections 6 and 7, the simulated and experimental results are compared in order to show the reliability of the complete solution found by the proposed algorithm.

METHODOLOGY

The methodology used in the definition and resolution of the algorithm is based on a model that departs from an initial solution, defined by the technological constraints that characterize the multistage deep-drawing processes (see Section 3). The use of these processes in mass production will require the minimization of the total process time and the quantity of raw material, thus acting as constraints.

The algorithm resolution has been carried out by the software called deep-drawing tool (DDT) developed by the author. This tool allows the user to perform the total multistage process; it allows the evaluation of the most important variables of the process and the implementation of the optimization algorithm which is the objective of this paper. The algorithm results are contrasted experimentally by means of the resolution of several industrial cases. The applied product type is an axisymmetric component produced in brass by multistage deep-drawing and ironing processes. The use of this material is justified due to its good properties suitable for these multistage processes. Table 1 shows the characteristics of brass UNS C26000 used on the resolution of the industrial cases presented in Section 5.

Table 1: Brass UNS C26000 properties

Material	UNS C26000
Density, ρ	8.53 Kg/dm^3
Material rigid-plastic constant, C	895.0 MPa
Strain hardening exponent, n	0.485
Tensile strength, yield, S_y	435.0 MPa
Tensile strength, ultimate, S_u	525.0 MPa
Normal anisotropy value, R	0.83

The solutions provided by this algorithm have been tested against three industrial cases and geometries as shown in Table 2.

Table 2: Dimensions of parts (in mm)

	Case A	Case B	Case C
External diameter, d_n	98.5	134.4	102.7
Length, l_n	620	650	1,480
Bottom thickness, s_n	13.5	13.8	15
Wall thickness, e_n	2	1.45	0.95

The resolution of these types of cases requires the combination of different processes of drawing, redrawing, and ironing. Figure 1 shows the drawing and redrawing/ironing press used on the manufacturing processes of the industrial cases. Figure 2 shows some of the manufacturing stages for Case A.

Figure 1: Redrawing and ironing press.

Figure 2: Case A: manufacturing stages.

MODEL APPROACH

Figure 3 shows the flow diagram of the model providing the initial solution: model phases, input variables required to define the "new project," and the flow information between different stages. The model develops the required solution from the input data. These input data correspond to the dimensions and material of the final piece. The required input data are the external diameter (d_n), the length (l_n), bottom thickness (s_n), wall thickness (e_n), and the material type. The model permits the selection of the material type and its mechanical characteristics needed to form the part: density, ρ (kg/m3); ultimate tensile strength, S_u (MPa); yield tensile strength, S_y (MPa); material rigid-plastic constant, C (MPa); strain hardening exponent, n; and normal anisotropy value, R. From the input data, the dimensions of the initial part or blank are calculated based on the incompressibility condition of the plastic deformation. This condition considers that the volume of the piece is unchanged throughout the deformation process [18]. A detailed formulation about the previous model that determines the initial solution can be found in Ramirez et al. [22]. The dimensions of this blank are the starting point to develop the drawing, redrawing, and ironing processes until the required final dimensions are achieved. In this way, data from the blank becomes the input for the next step: drawing.

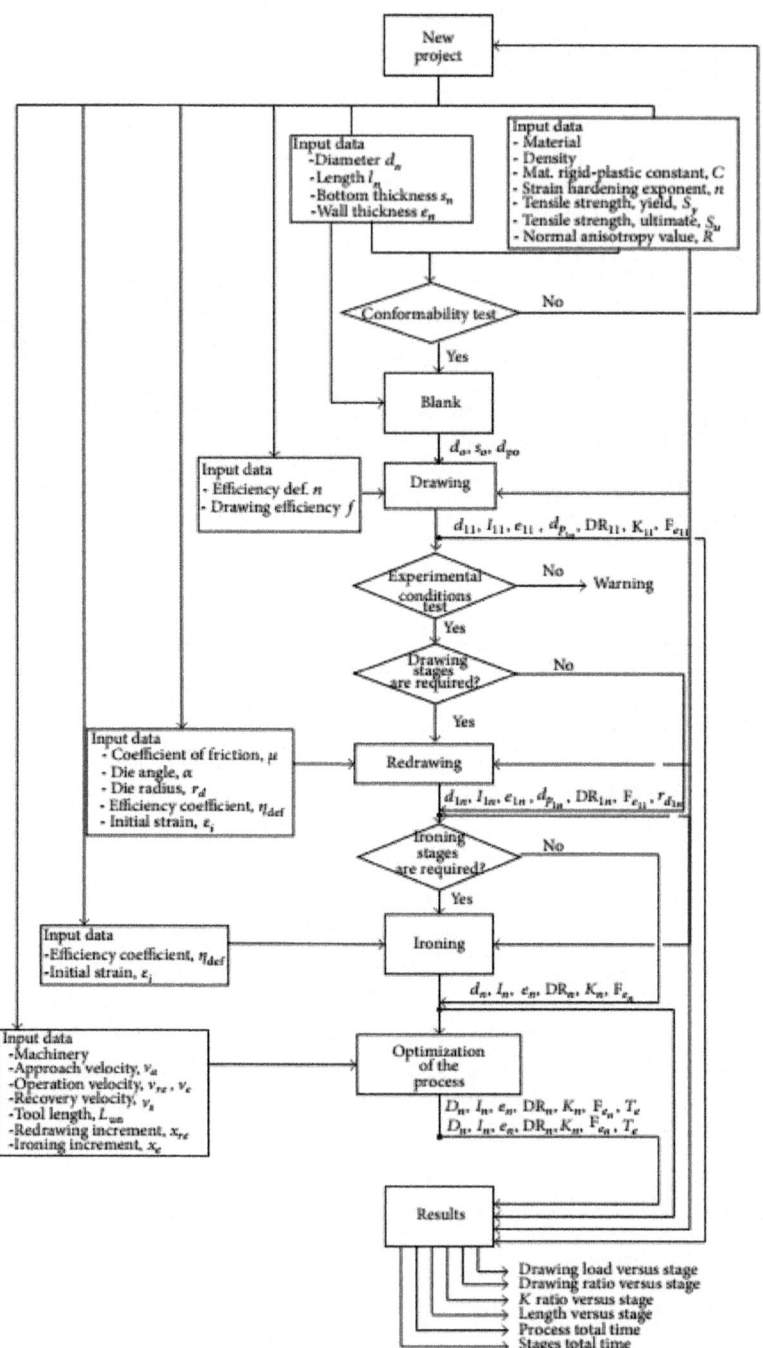

Figure 3: Flow diagram of the model.

Drawing

The model presents a calculation procedure for each process: drawing, redrawing, and ironing. The determination of the initial solution is treated independently for each of the threads that may occur. It is important to note that, depending on the type of piece, in some cases, its geometry is not necessary to consider redrawing or ironing stages. For determining the initial solution of the drawing, a hypothesis that would later be amended in the adjustment of the technological process is used, and it is the size of the diameter of the punch corresponding to that stage. The LDR is a measure of deep draw ability which defines the largest blank that can be drawn without tearing. It is the ratio between the maximum blank diameter and punch diameter. The design of this punch will depend on the LDR and the final dimensions of the piece. It is also possible that this diameter is changed later in the adjustment of the technological process in order to allow a clearance between the punch and the workpiece for subsequent stages of redrawing and ironing [20]. The solution for the initial stage of drawing is calculated by considering two limiting conditions as described below. The model selects the drawing diameter or the die, d, by using the largest diameter ($d_{1,1}$ and $d_{1,2}$) obtained from the two drawing conditions. Once the diameter is known, the model determines all the dimensions needed to define the drawing stage. Verification of the drawing stage dimension is conducted using two experimental conditions based on empirical studies and collected field data. The corresponding flow diagram is shown in Figure 4.

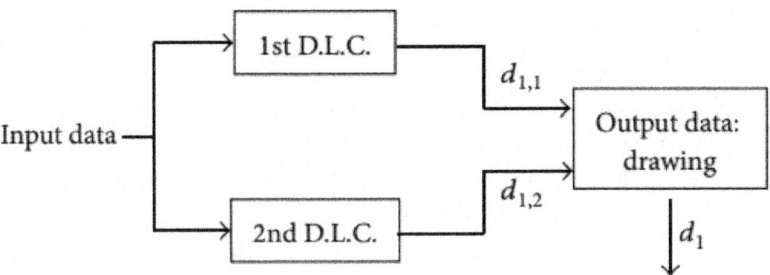

(D.L.C.: drawing limiting condition)

Figure 4: Flow diagram of the drawing algorithm.

The first limiting drawing condition is related to the maximum force ($F_{e,max}$) executed by the punch on the workpiece during the drawing process, which must be less than the cracking load of the material (F_{cr}), according to (1). These forces establish the range in which the force can be obtained using from the efficiency coefficient (η_{def}), ultimate tensile strength, blank thickness (s_0), and the initial (d_0) and final (d_1) diameters (note that the mean wall diameter is $d_{m,1}$ = $d_1 + s_0$), and they are based on the experimental expression from Siebel and Beisswanger [23]. This condition requires the following:

$$F_{e,max} = \pi \cdot d_{m,1} \cdot s_0 \left[1.1 \cdot \frac{1.3S_u}{\eta_{def}} \left(\ln \frac{d_o}{d_1} - 0.25 \right) \right],$$

$$F_{cr} = \pi \cdot d_{m,1} \cdot s_0 \cdot S_u.$$

$$(1)$$

The second drawing limiting condition is calculated by using the expression developed by Leu [1], which showed good agreement between theoretical and experimental results [2, 3]. Considering the condition of constant volume throughout the process of plastic deformation, can be defined as follows:

$$LDR = \sqrt{e^{(2fe^{-n\sqrt{(1+R)/2}})} + e^{(2n\sqrt{(1+R)/2})} - 1}.$$

$$(2)$$

This expression estimates the LDR as a function of normal anisotropy, the strain hardening exponent, and efficiency (f).In this manner, upper and lower bounds are established, considering the materials proprieties and the drawing capacity. The two limiting conditions use new data from the blank geometry and process efficiency (η_{def} and f).

Redrawing

Once the dimensions of the drawing stage are obtained, the model provides these data as input values for the next phase of the process: redrawing. The goal of this step is to obtain the final dimensions of the piece needed to perform the next step: ironing. The solution for the initial stage of redrawing is calculated from the consideration of three limiting conditions as described below. The model selects the diameter of the die from the largest of the solutions obtained in the three redrawing constraints ($d_1 \cdot n_{,1}$, $d_1 \cdot n_{,2}$, and $d_1 \cdot n_{,3}$). Once the diameter is known, the model determines all dimensions needed to define this stage. Figure 5 presents the flow diagram of this redrawing algorithm..

The first redrawing limiting condition is related to the redrawing load of the punch during the process of redrawing ($F_{re,max}$), which must be less than the cracking load of the material (F_{cr}), according to (3) and (4), respectively. Consider

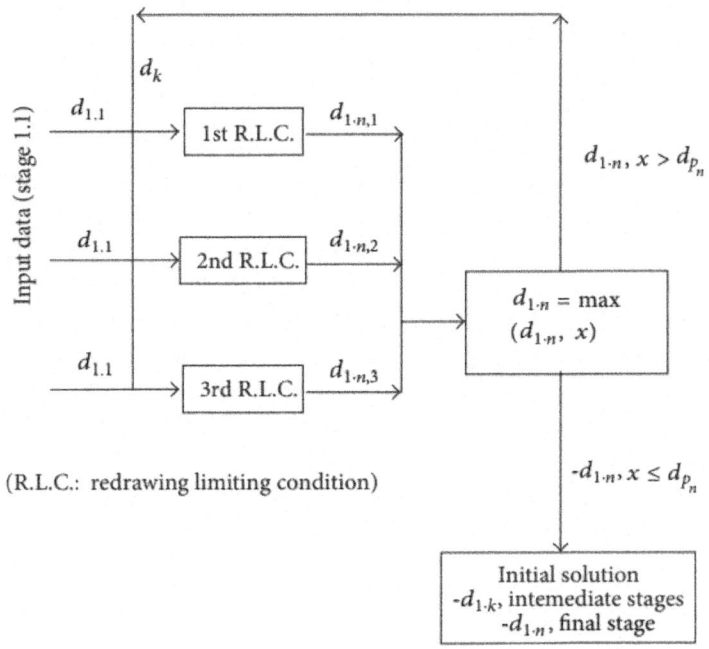

Figure 5: Flow diagram of the redrawing algorithm.

$$F_{re,max} = \pi \cdot d_{m,1 \cdot n} \cdot s_{1 \cdot n}$$

$$\cdot 1.3S_u \left[2e^{\mu\pi/2} \frac{d_{p,1 \cdot n} - d_{m,1 \cdot n}}{d_{p,1 \cdot n} + d_{m,1 \cdot n}} 1.1 + \frac{\mu}{\tan \alpha} + \frac{s_{1 \cdot n}}{2r_D} \right], \tag{3}$$

$$F_{cr} = \pi \cdot d_{m,1 \cdot n} \cdot s_{1 \cdot n} \cdot S_u. \tag{4}$$

The first condition is based on Siebel and Beisswanger " [23], where μ is the friction coefficient, α is the die half-angle, d_p is the punch diameter, and r_D is the die radius.

The second redrawing limiting condition assumes the rigid-plastic condition of the material. Assuming that the material behaves according to a parabolic law that approximates the potential behaviour of metallic materials coldmanufactured, then it is possible to determine an expression relating the external diameter of the part of a generic step of the process of redrawing, $d1 \cdot n$, to the diameter of the previous stage. Assuming the material is annealed ($\varepsilon_i = 0$) and considering the total deformation between the initial and final states, the condition takes the following form:

$$d_{1 \cdot n} = d_{1 \cdot n-1} e^{-\varepsilon_f/2}. \tag{5}$$

The third redrawing limiting condition is referred to as the limiting drawing ratio. In applying the third limiting condition of drawing, the model applies the formulation defined by Sonis et al. [2] about the LDR study in the operations of redrawing. The model considers the effects of the normal anisotropy of the material, friction coefficient, coefficient of strain hardening, and the radius of the input die (r_d). The LDR is used in this model as a variable to determine the required number of redrawing steps and size of the stages. It is assumed that the material is rigid-plastic [2]. Moreover, assuming that the material is rotationally symmetric, the same properties are based on the existence of normal anisotropy and planar isotropy. The Sonis model [2] is based on the tension that is created in the area of the radius of the die redrawing causing instability in the plastic wall of the cup, which is equal to the radial tension in the drawing area of the flange, due to the continuity of tension throughout the piece. Based on the Sonis model, the expression for LDR_i can be written as

$$
\begin{aligned}
& f(LDR_i) \\
&= -\frac{C_1 \cdot r_{c_{(i-1)}}}{r_{c_{(i-1)}} + LDR_{(i-1)} \cdot r_d} + \frac{C_3 \cdot r_{c_{(i-1)}}}{r_d + r_{c_{(i-1)}}/LDR_{(i-1)}} \\
& + C_4 \ln\left(\frac{r_{c_{(i-1)}}}{r_d + r_{c_{(i-1)}}/LDR_{(i-1)}} \right) \\
& + C_5 \ln\left(\frac{r_{c_{(i-1)}}}{r_d + r_{c_{(i-1)}}/LDR_{(i-1)}} \right) \\
& - \frac{1}{\left(r_d + r_{c_{(i-1)}}/LDR_{(i-1)}\right)^2 \left((R_{(i+1)}/r_{cd})^2 - 1\right)} \\
& \times \Bigg[r_{c_{(i-1)}} \\
& \quad \times \sqrt{\left(r_d + \frac{r_{c_{(i-1)}}}{LDR_{(i-1)}}\right)^2 \left(\left(\frac{R_{(i+1)}}{r_{cd}}\right)^2 - 1\right) + r_{c_{(i-1)}}^2} \\
& \quad - \left(r_d + \frac{r_{c_{(i-1)}}}{LDR_i}\right)^2 \left(\frac{R_{(i+1)}}{r_{cd}}\right) \\
& \quad - \left(r_{c_{(i-1)}}^2 - \left(r_d + \frac{r_{c_{(i-1)}}}{LDR_i}\right)\right)^2 \Bigg] \\
& + C_5 \ln\Bigg[r_{c_{(i-1)}} \\
& \quad + \sqrt{\left(r_d + \frac{r_{c_{(i-1)}}}{LDR_{(i-1)}}\right)^2 \left(\left(\frac{R_{(i+1)}}{r_{cd}}\right)^2 - 1\right) + r_{c_{(i-1)}}^2} \Bigg] \\
& - C_5 \ln\left(\left(\frac{r_d + r_{c(i-1)}}{LDR_i}\right)\left(1 + \frac{R_{(i+1)}}{r_{cd}}\right)\right) = 0, \tag{6}
\end{aligned}
$$

where C_1, C_3, C_4, and C_5 are constants, r_c is the corner radius, and r_{cd} is the die opening radius [2].

Based on the previous expressions in the function under the LDR, an expression depending on the LDR is obtained, and the nonlinear equations are solved by the NewtonRaphson method. In this way, it is possible to determine the values of LDR for each stage of redrawing, starting from an initial die radius and by the consideration that the die radius of each redrawing step will be reduced until 80% of the corresponding value of the previous phase. Once the LDR for each redrawing phase is found, the model determines the diameter of each stage.

Ironing

The number of stages of ironing depends on the size of the final part. The model is based on the performance of three limiting conditions to be drawn from stage 2 to stage n. The calculation of the process variable is the external diameter of the piece, according to the block diagram, illustrated in Figure 6.

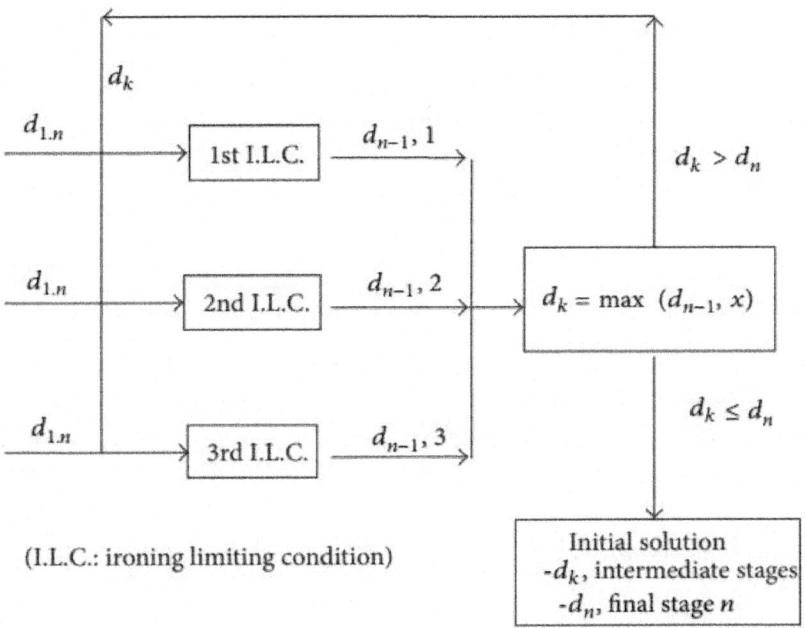

Figure 6: Flow diagram of the ironing algorithm.

By considering separately the merits of the three limiting conditions for ironing, the solution can be determined for phases 2 to n.

The maximum ironing load ($\sigma_{e,max}$) must be less than the cracking load (σ_r) of the material. This condition is expressed as the following

$$F_{e,n-1} < \frac{\pi}{4} \cdot \left(d_{n-2}^2 - d_{n-1}^2\right) \cdot n \cdot S_u.$$

(7)

Second Ironing Limiting Condition. This condition assumes that the maximum load in the ironing process (σ_e) must be less than the yield tensile strength ((ε_f)). This condition provides good results in parts of drawing [24] and it requires

$$\frac{C}{n+1}\left(\varepsilon_f^{n+1} - \varepsilon_i^{n+1}\right) < Y\left(\varepsilon_f\right) \equiv C \cdot \varepsilon_f^n.$$

(8)

Third Ironing Limiting Condition. This refers to the value of the thickness reduction ratio limit (K). This ratio is widely used in the calculation and design of the processes of drawing [5]. The model performs the calculation for each K stage, which mainly depends on the stage of drawing and the type of material used.

OPTIMIZATION PROCESS

The model is based on the minimization of the total process time, which is the time to approach the punch from the initial stage, plus the time of the operation, plus the time to recover the punch to its initial position, according to the following objective function:

$$F_o = \min t_e = \min \sum_{i=1}^{n} t_i = \min \sum_{i=1}^{n} \left(t_{a,i} + t_{o,i} + t_{s,i}\right),$$

(9)

where F_o is the objective function, te is the total process time, t_i is the total process time at stage i, $t_a i$ is the approach time at stage i, $t_{o,i}$ is the operation time at stage i, and ts,i is the recovery time at stage i. The process time at stage i is given by the expression

$$t_i = \frac{1.1 \cdot l_{i-1}}{V_{a,i}} + \frac{L_{ui} + l_i}{V_{e,i}} + \frac{1.1 \cdot l_{i-1} + L_{ui} + l_i}{V_{s,i}},$$

(10)

where l_{i-1} is the part length in stage $i-1$, Lui is the punch length in stage i, l_i is part length in stage i, $V_{a,i}$ is the approach velocity of the press in the stage i,

$V_{e,i}$ is the ironing velocity in the stage i, and $V_{s,i}$ is the recovery velocity of the press in the stage i.

where l_{i-1} is the part length in stage $i-1$, Lui is the punch length in stage i, $_{li}$ is part length in stage i, $V_{a,i}$ is the approach velocity of the press in the stage i, $V_{e,i}$ is the ironing velocity in the stage i, and $V_{s,i}$ is the recovery velocity of the press in the stage i.

$$F_o = \min t_e = \min \sum_{i=1}^{n} t_i$$

$$= \min \sum_{i=1}^{n} \left(\frac{1.1 \cdot l_{i-1}}{v_{a,i}} + \frac{L_{ui} + l_i}{v_{e,i}} + \frac{1.1 \cdot l_{i-1} + L_{ui} + l_i}{v_{s,i}} \right) \tag{11}$$

subject to

$$d_n \le D_n \le d_n (1 + \Delta DR_e), \tag{12}$$

where

$$l_i = \frac{s_i \left(d_o^2 - d_{p,i}^2 \right)}{\left(D_i^2 - d_{p,i}^2 \right)}, \tag{13}$$

and D_i is the optimized diameter in stage i, si is the bottom thickness in stage i, $d_{p,i}$ is the punch diameter in stage i, and Δ_{DRe} is the drawing surplus ratio between the stages n and m, defined by the expression

$$\Delta DR_e = DR_m - DR_n. \tag{14}$$

The optimization requires a distribution between the stages 1 to n of the drawing surplus ratio, ΔDR_e. This distribution is conditioned to the minimization of the total process time as the sum of each stage time. Thus, an improvement is achieved in each drawing stage. In this manner, the optimized drawing conditions are separate from the maximum limits fixed by the initial drawing conditions.

The resolution of the recursive function is carried out according to the values of all the possible diameters between d_n and $d(1 + \Delta DR_e)$. The algorithm identifies which combination is the most suitable, and it selects the optimized diameter D_n in each stage, such that the condition of the objective function is satisfied; that is, the combination of diameters must give the minimum total process time.

The algorithm also permits the modification of the velocity parameters in each press required in the multistage process (approach velocity, operation velocity, and recovery velocity of the punch). Accordingly, it is possible to realise a more adequate distribution of available presses in the facility.

ALGORITHM RESOLUTION PROCESS

The algorithm proposed in Figure 7 performs the resolution in the following steps.

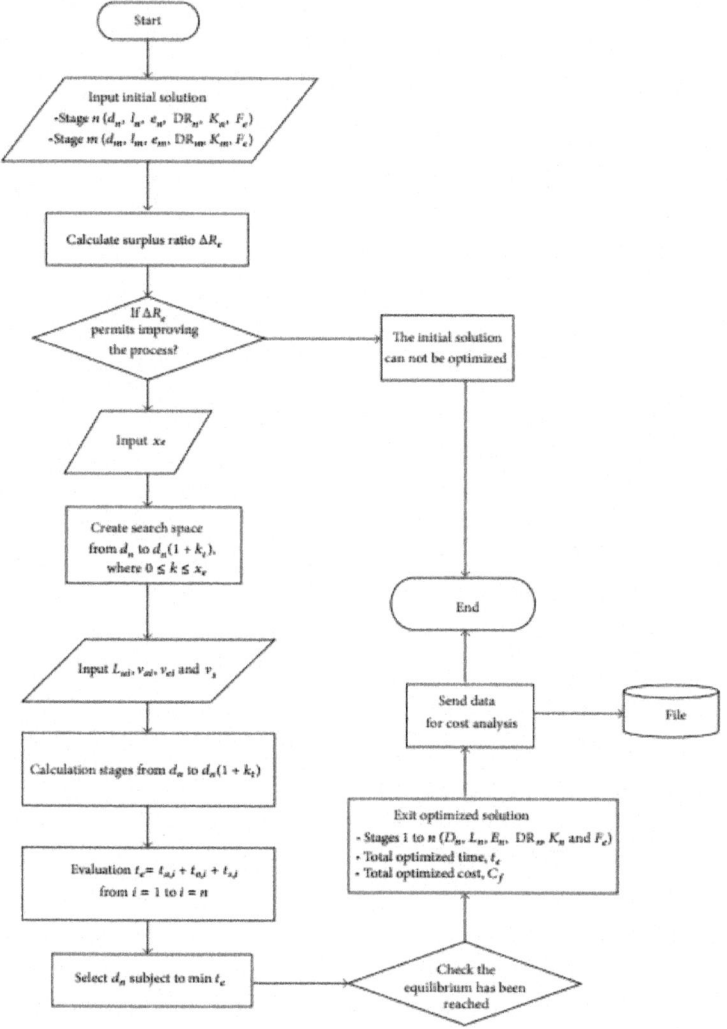

Figure 7: Flowchart of the algorithm.

Step 1. Definition of incremental factor t, given by the following expression:

$$t = \frac{\Delta DR_e}{x_e},$$

(15)

where ΔDR_e is the surplus ratio of drawing and x_e is the number of times that the drawing surplus ratios are fractioned.

Step 2. Progressive increment of diameter of each process stage, from d_n to $d(1 + k_t)$, where k is a parameter that varies from 0 to factor x_e.

Step 3. Resolution of the recursive function is based on obtaining all possible substages, by means of the modification of each stage diameters, from d_n to $d_n(1 + \Delta DR_e)$.

Step 4. Once all the possible diameters of each stage are defined, the algorithm searches the arrangement that allows for the minimizing of the total process time.

Step 5. The search stops and the best solution up to the current iteration is given as the output.

EXPERIMENTAL RESULTS

The deep-drawing tool has allowed for the verification of this algorithm's integrity. Computational and experimental tests have been carried out in brass, in particular in UNS C26000 alloy, applied to three parts. The dimensions of the parts are shown in Table 2.

For the experimental results of the three cases presented in Table 2, its analytical resolution has been carried out by the software DDT according to the flowchart presented in Figure 3. The final result of using the software tool is an improved solution that optimizes the total process time and reduces the manufacturing cost of the presented industrial cases. These costs use the information from the results provided by the previous steps and are based on the total time of the deformation process and the labour cost per hour.

Figure 8 shows the results of the tool for the initial solution corresponding to Case A.

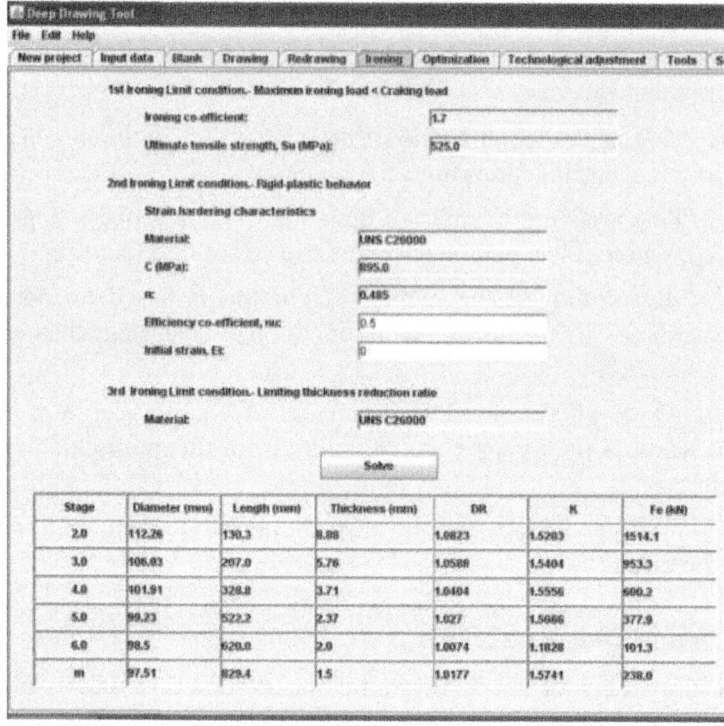

Figure 8: Case A: initial solution results.

Figure 9 presents the calculation of the optimized process by means of the DDT using the proposed algorithm. Values for x_{re} and x_e in operations of redrawing and ironing are given.

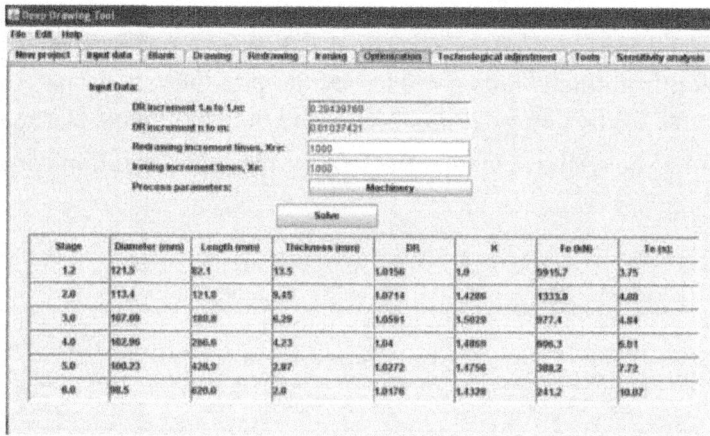

Figure 9: Case A: optimization process results.

The resolution of the optimization process using the proposed algorithm requires the introduction of the machinery operation parameters involved in the process.

The computer tool requires the input of these data (Figure 10), according to the flowchart of the algorithm presented in Figure 3.

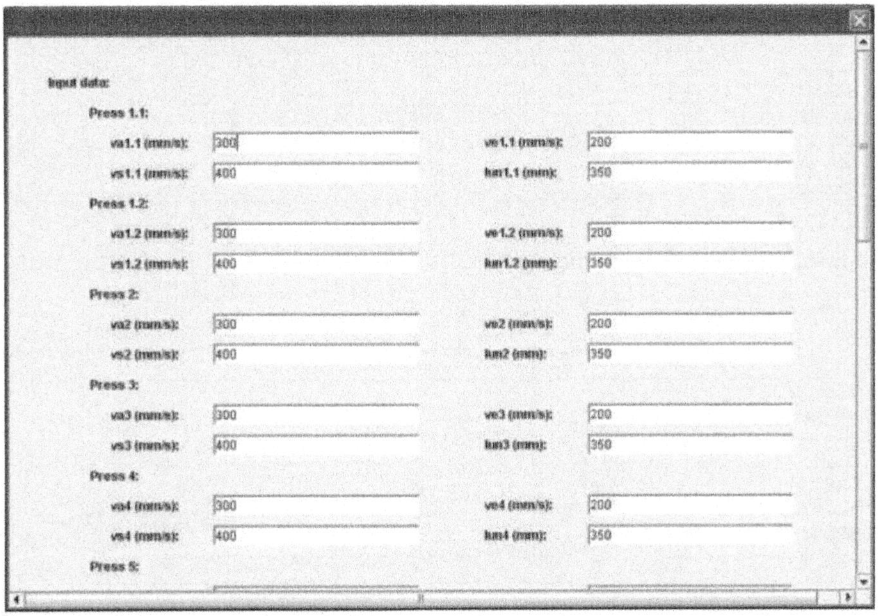

Figure 10: Machinery parameters.

One of the most important variables in the analysis of such processes is the variable "ratio of wall thickness reduction." The optimization process allows the user to obtain a greater stability of the process maintaining the balance of the different phases. Table 3 shows the evolution of the drawing load for the industrial Cases A, B, and C under study. Table 4 shows the evolution of the wall thickness. The results show that the optimized process achieves a better balance so that it is more stable. This stability is transmitted in a better distribution of the capabilities of drawing, redrawing, and ironing between the different stages, subsequently resulting in improved processing time and reduced manufacturing cost.

Table 3: Evolution of drawing load

Stages	Case A		Case B		Case C	
	Initial solution (IS)	DDT algorithm solution	Initial solution (IS)	DDT algorithm solution	Initial solution (IS)	DDT algorithm solution
Drawing						
Stage 1.1	3188.5	3188.5	3175.0	3175.0	3692.1	3692.1
Redrawing						
Stage 1.2	5915.7	5915.7	—	—	—	—
Ironing						
Stage 2.0	1514.1	1346.4	2082.3	1859.8	1803.8	1710.1
Stage 3.0	953.3	962.6	1311.1	1320.5	1135.7	1141.1
Stage 4.0	600.2	626.8	825.5	859.3	715.1	718.6
Stage 5.0	377.9	381.4	519.8	523.2	450.2	452.6
Stage 6.0	101.3	229.8	327.3	329.5	283.5	285.0
Stage 7.0			15.8	189.3	178.5	179.4
Stage 8.0					32.4	112.4

Table 4: Evolution of wall thickness

Stages	Case A		Case B		Case C	
	Initial solution (IS)	DDT algorithm solution	Initial solution (IS)	DDT algorithm solution	Initial solution (IS)	DDT algorithm solution
Drawing						
Stage 1.1	1	1	1	1	1	1
Redrawing						
Stage 1.2	1	1	—	—	—	—
Ironing						
Stage 2.0	1.520	1.434	1.535	1.450	1.518	1.477
Stage 3.0	1.540	1.494	1.552	1.505	1.538	1.516
Stage 4.0	1.555	1.511	1.564	1.519	1.554	1.520
Stage 5.0	1.566	1.474	1.572	1.479	1.565	1.512
Stage 6.0	1.182	1.412	1.578	1.442	1.573	1.492
Stage 7.0			1.028	1.345	1.578	1.459
Stage 8.0					1.11	1.409

Once the evolution of different variables that influence the process has been analyzed, the results for "process time" and "manufacturing cost" are presented in order to study the improvements that could be achieved by the implementation of the proposed algorithm.

Tables 5 and 6 show the evolution of these improvements on the variables "total process time" (t_e) and "manufacturing cost" (C_m) in terms of number of iterations performed by the algorithm

Table 5: Process time evolution as a function of the algorithm iteration number

x_e	5	25	50	100	250	500	1000
$t_{e,1}$	42.054	42.054	42.054	42.054	42.054	42.054	42.054
$t_{e,n}$	40.634	40.109	39.998	39.952	39.889	39.813	39.797
% reduction	3.67%	5.02%	5.31%	5.43%	5.59%	5.79%	5.83%

Table 6: Manufacturing cost evolution as a function of the algorithm iteration number

x_e	5	25	50	100	250	500	1000
Cm_1	34.974	34.974	34.974	34.974	34.974	34.974	34.974
Cm_n	33.554	33.029	32.918	32.872	32.809	32.733	32.717
% reduction	4.06%	5.56%	5.88%	6.01%	6.19%	6.41%	6.45%

Similarly, the evolution of "total process time" and the "manufacturing cost" for Cases B and C has been analyzed. Figures 11 and 12 show the improvements evolution obtained for the three Cases A, B, and C. It can be observed that the greater improvements are achieved in the first iterations (first five according to Case A), but the results improve further as the iteration increases.

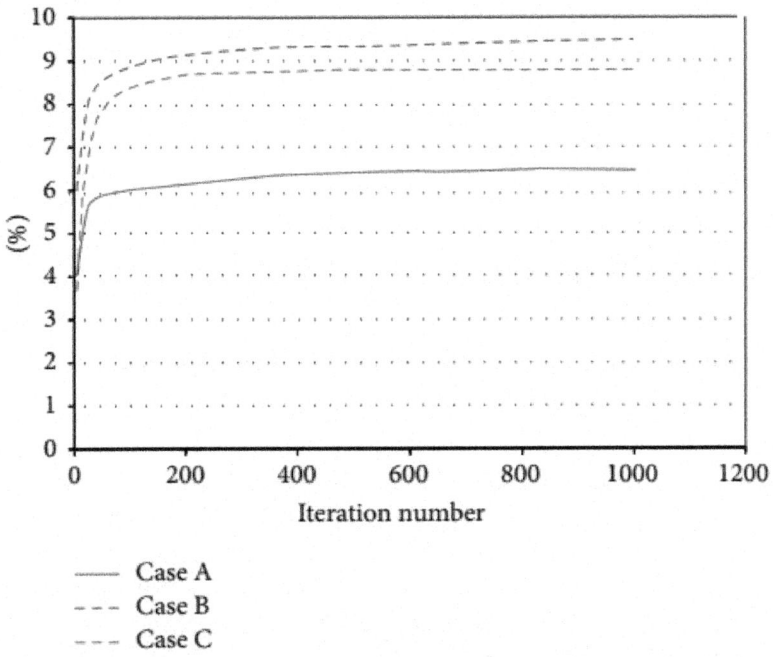

—— Case A
--- Case B
--- Case C

Figure 11: Cases A, B, and C: process time improvements as function of the algorithm iteration number.

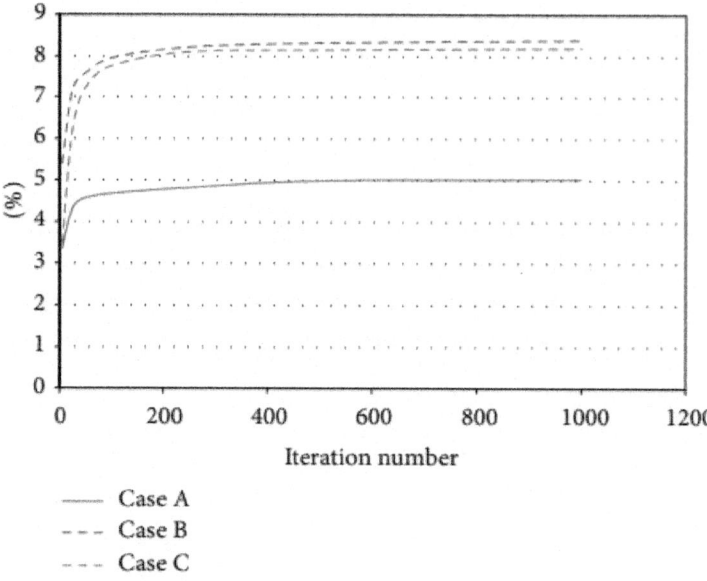

Figure 12: Cases A, B, and C: manufacturing cost improvements as function of the algorithm iteration number.

ANALYSIS AND DISCUSSION OF THE RESULTS

The resolution of the industrial Cases A, B, and C by means of the DDT Algorithm presents important advantages. As shown in Tables 7, 8, and 9, the DDT algorithm produces improvements between 5.17% and 8.18% compared with the initial solution for the process time variable and between 4.40% and 7.78% for the manufacturing cost. In the same manner, the experimental results obtained an improvement between 6.55% and 9.34% for the process time and from 6.60% to 11.55% for the manufacturing cost. These improvements are more relevant in mass production. The most significant advances have been achieved in Case B, which represent the highest progress according to the algorithm iteration number (see Figures 11 and 12). Also there is similarity between this algorithm iteration number and Cases A and C where the trend is the same in all cases. The final geometry of the part does not seem to have a particular effect on the results. This is due to a good definition of initial solution that allows processing the blank.

Table 7: Case A: analysis of the results

Comparative analysis	Process time	Manufacturing cost per unit
Initial solution (IS)	34.83 s	0.318€
DDT algorithm solution	33.03 s	0.304€
Experimental solution (EXS)	32.55 s	0.297€
DDT versus IS improvement	5.17%	4.40%
EXS versus IS improvement	6.55%	6.60%
EXS versus DDT accuracy	98.55%	97.7%

Table 8: Case B: analysis of the results.

Comparative analysis	Process time	Manufacturing cost per unit
Initial solution (IS)	42.4 s	0.398€
DDT algorithm solution	38.93 s	0.367€
Experimental solution (EXS)	38.44 s	0.352€
DDT versus IS improvement	8.18%	7.78%
EXS versus IS improvement	9.34%	11.55%
EXS versus DDT accuracy	98.74%	95.92%

Table 9: Case C: analysis of the results

Comparative analysis	Process time	Manufacturing cost per unit
Initial solution (IS)	76.58 s	0.688€
DDT algorithm solution	70.95 s	0.639€
Experimental solution (EXS)	69.82 s	0.627€
DDT versus IS improvement	7.35%	7.12%
EXS versus IS improvement	8.83%	8.88%
EXS versus DDT accuracy	98.41%	98.13%

Comparing the algorithm results with the experimental results shows that the accuracy of the process is very high, obtaining values from 98.41% to 98.74% for the process time variable and from 95.92% to 98.13% for the manufacturing cost. Although there are some disparities between the cases analyzed, these differences are not significant because the accuracy is always above 95% and normally very close to 99%. Thus, the mathematical minimization of the total process time through the technological variables provides a solution which is very close to the experimental outcomes.

CONCLUSIONS

In this paper, an algorithm that allows the reduction of the total process times and cost in the manufacturing of axisymmetric components has been presented. The algorithm is based on the minimization of total process time, defined by means of the part dimensions and process velocities, and costs by considering the reduction of material usage, through constraints related to drawing surplus ratio. The algorithm has been enhanced with the use of technological parameters. The simulation results of the algorithm achieved a good agreement with the experimental results obtained for the three cases and caused significant improvements in manufacturing times and costs in the deep-drawing process of axisymmetric parts. The comparison of analytical results obtained with the experimental results has demonstrated the high accuracy of the algorithm, which is of interest for real industrial applications.

Future works will aim to get a more efficient process, from a perspective of sustainable energy, thus achieving an integral solution, in terms of scientific and technological basis.

ACKNOWLEDGMENTS

The authors gratefully acknowledge the company EXPAL Systems, S.A., as well as its factory in Navalmoral de la Mata (Cáceres, Spain) for supporting the experimental results presented in this paper.

REFERENCES

1.   D. K. Leu, "The limiting drawing ratio for plastic instability of the cup-drawing process," Journal of Materials Processing Technology, vol. 86, no. 1–3, pp. 168–176, 1998.

2.   P. Sonis, N. V. Reddy, and G. K. Lal, "On multistage deep drawing of axisymmetric components," Journal of Manufacturing Science and Engineering, vol. 125, no. 2, pp. 352–362, 2003.

3.   R. K. Verma and S. Chandra, "An improved model for predicting limiting drawing ratio," Journal of Materials Processing Technology, vol. 172, no. 2, pp. 218–224, 2006.

4.   R. Padmanabhan, M. C. Oliveira, A. J. Baptista, J. L. Alves, and L. F. Menezes, "Blank design for deep drawn parts using parametric NURBS

surfaces," Journal of Materials Processing Technology, vol. 209, no. 5, pp. 2402–2411, 2009.

5. Agrawal, N. V. Reddy, and P. M. Dixit, "Determination of optimum process parameters for wrinkle free products in deep drawing process," Journal of Materials Processing Technology, vol. 191, no. 1–3, pp. 51–54, 2007.

6. M. Haddadzadeh, M. R. Razfar, and M. R. M. Mamaghani, "Novel approach to initial blank design in deep drawing using artificial neural network," Proceedings of the Institution of Mechanical Engineers B: Journal of Engineering Manufacture, vol. 223, no. 10, pp. 1323–1330, 2009.

7. Chamekh, S. Ben Rhaiem, H. Khaterchi, H. Bel Hadj Salah, and R. Hambli, "An optimization strategy based on a metamodel applied for the prediction of the initial blank shape in a deep drawing process,"International Journal of Advanced Manufacturing Technology, vol. 50, no. 1–4, pp. 93–100, 2010.

8. S. Sezek, V. Savas, and B. Aksakal, "Effect of die radius on blank holder force and drawing ratio: A model and experimental investigation," Materials and Manufacturing Processes, vol. 25, no. 7, pp. 557–564, 2010. · ·

9. R. Padmanabhan, M. C. Oliveira, J. L. Alves, and L. F. Menezes, "Influence of process parameters on the deep drawing of stainless steel," Finite Elements in Analysis and Design, vol. 43, no. 14, pp. 1062–1067, 2007.

10. K. Mori and H. Tsuji, "Cold deep drawing of commercial magnesium alloy sheets," CIRP Annals-Manufacturing Technology, vol. 56, no. 1, pp. 285–288, 2007.

11. L. M. A. Hezam, M. A. Hassan, I. M. Hassab-Allah, and M. G. El-Sebaie, "Development of a new process for producing deep square cups through conical dies," International Journal of Machine Tools and Manufacture, vol. 49, no. 10, pp. 773–780, 2009.

12. H. C. Tseng, C. Hung, and C. C. Huang, "An analysis of the formability of aluminum/copper clad metals with different thicknesses by the finite element method and experiment," International Journal of Advanced Manufacturing Technology, vol. 49, no. 9–12, pp. 1029–1036, 2010.

13. B. C. Hwang, S. M. Han, W. B. Bae, and C. Kim, "Development of an automated progressive design system with multiple processes (piercing, bending, and deep drawing) for manufacturing products,"International Journal of Advanced Manufacturing Technology, vol. 43, no. 7-8, pp. 644–653, 2009.

14. K. V. Ramana and P. V. M. Rao, "Data and knowledge modeling for design-process planning integration of sheet metal components," Journal of Intelligent Manufacturing, vol. 15, no. 5, pp. 607–623, 2004.

15. G. C. Vosniakos, I. Segredou, and T. Giannakakis, "Logic programming for process planning in the domain of sheet metal forming with progressive dies," Journal of Intelligent Manufacturing, vol. 16, no. 4-5, pp. 479–497, 2005.

16. F. Javier Ramírez and R. Domingo, "Application of an aided system to multi-step deep drawing process in the brass pieces manufacturing," in Proceedings of the 3rd Manufacturing Engineering Society International Conference, MESIC 2009, pp. 370–379, Alcoy, Spain, June 2009.

17. F. J. Ramirez, R. Domingo, and M. A. Sebastian, "Design of an aided system to optimise times and costs in deep drawing process," in Proceedings of the 2nd IPROMS International Researchers Symposium, vol. 1, pp. 191–196, Ischia, Italy, 2009.

18. K. Lange, Handbook of Metal Forming, McGraw-Hill, New York, NY, USA, 1985.

19. S. Y. Chung and S. H. Swift, "An experimental investigation into the re-drawing of cylindrical shells,"Proceedings of the Institution of Mechanical Engineers B: Journal of Engineering Manufacture, vol. 1, pp. 437–447, 1975.

20. M. A. Sebastián and A. M. Sanchez-Perez, "Diseño asistido por ordenador de los útiles para la embutición profunda de piezas cilíndricas huecas," Internal Report, ETSII, UPM, Madrid, Spain, 1980.

21. F. J. Ramirez, R. Domingo, M. A. Sebastian, and M. S. Packianather, "The development of competencies in manufacturing engineering by means of a deep-drawing tool," Journal of Intelligent Manufacturing, vol. 24, no. 3, pp. 457–472, 2011. ·

22. F. J. Ramirez, R. Domingo, and M. A. Sebastian, "A technological model applied to multi-stage deep drawing process of axisymmetric components," in Proceedings of the 21st International Computer-Aided Production Engineering Conference (CAPE ‹10), Edinburgh, UK, 2010.

23. E. Siebel and H. Beisswänger, Deep Drawing, Carl Hanser, Munich, Germany, 1995.

24. E. M. Rubio, M. Marín, R. Domingo, and M. A. Sebastián, "Analysis of plate drawing processes by the upper bound method using theoretical work-hardening materials," International Journal of Advanced Manufacturing Technology, vol. 40, no. 3-4, pp. 261–269, 2009.

CITATION

CHAPTER 1

Sylvestre Uwizeyemungu, Placide Poba-Nzaou, Josée St-Pierre, Assimilation Patterns in the Use of Advanced Manufacturing Technologies in Smes: Exploring Their Effects on Product Innovation Performance, DOI: 10.4301/S1807-17752015000200005

CHAPTER 2

S. Saberi, R. Mohd. Yusuff, N. Zulkifli and M.M.H. Megat Ahmad, 2010. Effective Factors on Advanced Manufacturing Technology Implementation Performance: A Review. *Journal of Applied Sciences, 10: 1229-1242*

CHAPTER 3

Ki-Hong Lee, In-Sung Yeo, Benjamin M. Wu, et al., "Effects of Computer-Aided Manufacturing Technology on Precision of Clinical Metal-Free Restorations," BioMed Research International, vol. 2015, Article ID 619027, 5 pages, 2015. doi:10.1155/2015/619027

CHAPTER 4

Kaufui V. Wong and Aldo Hernandez, "A Review of Additive Manufacturing," ISRN Mechanical Engineering, vol. 2012, Article ID 208760, 10 pages, 2012. doi:10.5402/2012/208760

CHAPTER 5

V. Schirosi, G. Del Re, L. Ferrari, P. Caliandro, L. Rizzi, and G. Melone, "A Novel Manufacturing Technology for RF MEMS Devices on Ceramic Substrates," Journal of Sensors, vol. 2010, Article ID 625325, 6 pages, 2010. doi:10.1155/2010/625325

CHAPTER 6

Samuel Kenzari, David Bonina, Jean Marie Dubois, and Vincent Fournée, Complex metallic alloys as new materials for additive manufacturing, doi:10.1088/1468-6996/15/2/024802

CHAPTER 7

Linan Zhu, Yanwei Zhao, and Wanliang Wang, "A Bilayer Resource Model for Cloud Manufacturing Services," Mathematical Problems in Engineering, vol. 2013, Article ID 607582, 10 pages, 2013. doi:10.1155/2013/607582

CHAPTER 8

Rajeev Saha and Sandeep Grover, "Identifying Enablers of E-Manufacturing," ISRN Mechanical Engineering, vol. 2011, Article ID 193124, 6 pages, 2011. doi:10.5402/2011/193124

CHAPTER 9

Alfred T. Sidambe, Biocompatibility of Advanced Manufactured Titanium Implants—A Review, doi:10.3390/ma7128168

CHAPTER 10

Hongyi Sun (2012). Linking Process Technology and Manufacturing Performance Under the Framework of Manufacturing Strategy, Management of Technological Innovation in Developing and Developed Countries, Dr. HongYi Sun (Ed.), ISBN: 978-953-51-0365-3, InTech, DOI: 10.5772/38557.

CHAPTER 11

Josef Hynek and Vaclav Janecek (2010). Advanced Manufacturing Technology Projects Justification, Mechatronic Systems Applications, Annalisa Milella Donato Di Paola and Grazia Cicirelli (Ed.), ISBN: 978-953-307-040-7, InTech, DOI: 10.5772/8926.

CHAPTER 12

Carlos Mireles, Alfonso Noriega and Gerardo Leyva (2006). Flexible Manufacturing System Simulation Using Petri Nets, Manufacturing the Future, Vedran Kordic, Aleksandar Lazinica and Munir Merdan (Ed.), ISBN: 3-86611-198-3, InTech, DOI: 10.5772/5048.

CHAPTER 13

F. Javier Ramírez, Rosario Domingo, Michael S. Packianather, and Miguel A. Sebastian, "Enhancing Multistage Deep-Drawing and Ironing Manufacturing Processes of Axisymmetric Components: Analysis and Experimentation," International Journal of Manufacturing Engineering, vol. 2014, Article ID 596128, 12 pages, 2014. doi:10.1155/2014/596128.

INDEX

A

Acrylonitrile butadiene styrene (ABS) 70
Additive manufacturing (AM) 63
Advanced manufacturing technologies (AMT) 1, 2, 10, 14, 15, 187
Advanced manufacturing technology (AMT) 23, 46, 142, 211
Automated guided vehicles (AGV) 144
Automated guided vehicles (AGVS) 27

B

Biocompatibility of the materials 153

C

Cloud emander (CD) 123
Cloud end (CE) 123
Cloud manufacturing service model (CMSM) 123
Cloud provider (CP) 123
Complex metallic alloys (CMAs) 101

Computer aided design (CAD) 187
Computer-aided design (CAD) 25, 63, 65, 99, 158, 159, 171
Computer-aided design/computer-aided manufacturing (CAD/CAM) 51, 52
Computer-aided digital technologies 58
Computer-aided drawing (CAD) 6
Computer-aided manufacturing (CAM) 12, 65
Computer-aided process planning (CAPP) 27
Computer integrated manufacturing (CIM) 215
Computer integrated manufacturing (CIM). 184
Computer numerical control (CNC) 144
Connectivity information 240
Customer relationship management (CRM) 145

D

Deep-drawing tool (DDT) 249
Distributed real-time embedded sys-

tems (DRES) 147

E

Electron beam melting (EBM) 65,
    72, 159, 178
Electron beam sintering (EBS) 159
Enterprise resource planning (ERP)
    144

F

Fabrication industry 153
Flexible manufacturing systems
    (FMS) 6
Food and drug administration (FDA)
    154, 158
Fused deposition modeling (FDM)
    65

I

Information technologies (IT) 2
Internal rate of return (IRR) 213,
    221
International standards organisation
    (ISO) 154

L

Laminated engineered net shaping
    (LENS) 65
Laminated Object manufacturing
    (LOM) 73
Large enterprises (LE) 2
Literature contributions 248

M

Manufacturing process 87, 90
Matrix provides 240
Microelectromechanical systems 85

N

Net present value (NPV), 212

O

Office automation (OA). 144
Operator inputs 243
Optimization requires 259
Optimization technique 247
Optimized diameter 259

P

Petri Nets (PN) 233
Polycarbonate (PC) 70
Programmable Logic Controllers
    (PLC) 233

R

Resource-based view (RBV) 16

S

Sacrificial layer 87, 88, 91
Scanning electron microscopy
    (SEM) 103
Selective laser sintering (SLS) 65,
    100
Semiconductor 86, 87
Silicone impression 52, 53, 54
Small and medium-sized enterprises
    (SMEs) 1, 2, 16
Society of manufacturing 63, 79, 80
Society of manufacturing engineer
    (SME) 187
Standard deviation (SD) 56
Stereolithography contour (SLC) 70
Supply chain management (SCM)
    148

T

Total quality management (TQM)
    183
Trigger a Transition 236